EMERGING TRANSACTIVE ENERGY TECHNOLOGY FOR FUTURE MODERN ENERGY NETWORKS

EMERGING TRANSACTIVE ENERGY TECHNOLOGY FOR FUTURE MODERN ENERGY NETWORKS

MOHAMMADREZA DANESHVAR
Researcher, Faculty of Electrical and Computer Engineering, University of Tabriz, Tabriz, Iran

BEHNAM MOHAMMADI-IVATLOO
Professor, Faculty of Electrical and Computer Engineering, University of Tabriz, Tabriz, Iran

KAZEM ZARE
Professor, Faculty of Electrical and Computer Engineering, University of Tabriz, Tabriz, Iran

ELSEVIER

ACADEMIC PRESS
An imprint of Elsevier
elsevier.com/books-and-journals

Academic Press is an imprint of Elsevier
125 London Wall, London EC2Y 5AS, United Kingdom
525 B Street, Suite 1650, San Diego, CA 92101, United States
50 Hampshire Street, 5th Floor, Cambridge, MA 02139, United States
The Boulevard, Langford Lane, Kidlington, Oxford OX5 1GB, United Kingdom

ISBN: 978-0-323-91133-7

For Information on all Academic Press publications visit our website at https://www.elsevier.com/books-and-journals

Publisher: Joseph P. Hayton
Acquisitions Editor: Rachel Pomery
Editorial Project Manager: Aera F Gariguez
Production Project Manager: Sreejith Viswanathan
Cover Designer: Miles Hitchen

Typeset by Aptara, New Delhi, India

Contents

Preface

Nowadays, energy networks are developing toward fully equipped with renewable energy resources due to the economic and environmental aspects along with significant problems of conventional units. However, several basic challenges are appeared regarding the reliable and secure operation of energy networks due to the uncontrollable outputs of renewable energy resources, the volatility of energy market prices, and stochastic behaviors of consumers under the high share of renewables as the very ambitious targets in electricity and/or the energy system. Additionally, the intermittent nature of renewable energy resources has led to substantial challenges regarding the integration of renewable energy resources throughout the energy grid, which needs advanced technologies for handling the large uncertainties and complexity of the system with a high penetration of renewable energy resources. To overcome such challenges, grid modernization is proposed as an effective solution for delivering greater quantities of zero-to low-carbon energy securely and reliably, including handling different types of renewable energy resources like solar and wind power. Transactive energy as an emerging technology plays a vital role in realizing modern multi-carrier energy networks with 100% renewable energy resources by focusing on developing local energy trading markets as well as significant economic and control mechanisms. Given the critical role of transactive energy technology in modernizing multi-carrier energy networks, this book aims to explore the modeling, optimization, and analyzing the necessity of the transactive energy technology for future modern energy networks, for an audience of energy, power, mechanical, chemical, process, and environmental engineers as well as the researchers and postgraduate students who work in the field of various types of energy systems. Indeed, this book concentrates on evaluating the capability of transactive energy in solving the challenges ahead of future modern multi-carrier energy networks. Moreover, it addresses how future modern energy networks can be reliably integrated and operated with 100% renewable energy resources under the transactive energy architecture.

The current book consists of eight chapters. Chapter 1 presents the definition of modern grids and the challenges, opportunities, principles, and objectives for grid modernization. Chapter 2 provides a comprehensive analysis for identifying the basic requirements for switching from current energy grids to modern energy networks. Moreover, it evaluates the potential of multi-carrier energy networks from the modernization stand of view by focusing on different energy networks. Chapter 3 discusses the potential of the peer-to-peer energy trading strategy in facilitating reliable switching from traditional uni-dimension networks to future modern multi-carrier energy grids. Furthermore, it examines the necessity for sustainable energy exchanging technologies based on the features of the peer-to-peer energy trading strategy. Chapter 4 covers the definition, framework, challenges, opportunities, principles, and strategies of transactive energy technology. It also investigates the capability of transactive energy technology in modernizing future multi-carrier energy networks. Chapter 5 includes the different applications of transactive energy technology in modernizing future hybrid energy networks. Besides, it discusses the impact of transactive energy on facilitating the presence of the new state of art technologies in the modern structure of the grid. Chapter 6 investigates the optimization of design, operation, and scheduling of community-based microgrids under the transactive energy paradigm. Moreover, it elaborates on designing the local multi-carrier energy trading environment in the interconnected structure of hybrid energy networks. Chapter 7 includes a case study to examine the capability of transactive energy technology in the reliable operation of the hybrid energy network with 100% renewable energy resources. It also models uncertainties of renewable-based systems as well as introduces the related tools and methods. Chapter 8 describes the application of GAMS software, its features, terms, symbols, and reserved words as well as the mathematical functions and commands completely. Also, it examines the application of different parts in GAMS by providing examples in line with hybrid energy networks.

As any research achievement may not be free of gaps, the Authors kindly welcome any suggestions and comments from the respectful readers for improving the quality of this book. Interested readers can share their valuable comments with the Authors via m.r.daneshvar95@gmail.com.

Mohammadreza Daneshvar
Faculty of Electrical and Computer Engineering,
University of Tabriz, Tabriz, Iran

Behnam Mohammadi-Ivatloo
Faculty of Electrical and Computer Engineering,
University of Tabriz, Tabriz, Iran

Kazem Zare
Faculty of Electrical and Computer Engineering,
University of Tabriz, Tabriz, Iran

CHAPTER

1

Overview of the grid modernization

1.1 Introduction

Nowadays, as the electrolization and digitization have become popular evolutions among people, forming developments movements for smart energy systems are known as prevalent trends. This evolution has raised the energy consumption rate in its different carriers such as electricity, heat, and gas. However, the consumption rate for multi-energy is expected to experience more growing changes by emerging new hybrid energy systems [1]. What is the key regarding such changes is to how the system can properly deal with this rising rate? To suitably respond, the question needs to be analyzed from different perspectives. From one viewpoint, new conventional power plants can be built for increasing the energy production capacity under the centralized energy production paradigm [2]. As this strategy enables the system for controlling energy generation, the operation of these units imposes extreme investment and operation costs for the system due to burning fossil fuels while releasing a huge amount of greenhouse gas emissions [3]. This strategy has been refused due to the aforementioned unacceptable disadvantages. From another viewpoint, mushrooming diverse renewable energy generation systems (REGSs) is warmly welcomed due to their economic and environmental benefits. However, the unpredictability features of REGSs have made the secure exploitation of the power grid challenge [4]. Another challenge stands on the growing appearance of hybrid systems that are developed for upsurging synergies among the different energy grids [5]. This issue inspires the fact of integrability of energy networks together. To sum up, energy networks need to operate a high/full REGSs for serving the

Emerging Transactive Energy Technology for Future Modern Energy Networks
DOI: https://doi.org/10.1016/B978-0-323-91133-7.00003-X

energy consumption of their consumers from clean energy sources considering their priorities, use capable technologies for reliably conducting this mission considering a high level of intermittences in the system, customize appropriate energy architecture for the interconnected version of the energy network's infrastructure, and use intelligent devices for establishing autonomous interactions between smart energy systems [6]. As the structure of current independent energy networks hinders achieving the aforementioned benefits for the grid, grid modernization is projected for multi-energy networks for the transition from the current system to modern multi-energy grids. Indeed, grid modernization is recognized as a catch-all phrase in referring to changes mentioned in previous sentences that accommodate all the rapid technological evolutions encompassing various parts of multi-energy grids from the energy generation sector to the energy distribution. In particular, grid modernization implies the growing application of information technologies, smart sensors and communication platforms, computers, etc., for making the structure of the energy grid more flexible than ever before. Therefore, this chapter is targeted to investigate future modern multi-energy grids and their features, principles, benefits, and challenges as well as scrutinize the whole range of adopting new emerging technologies that make the implementation of modern grids feasible.

1.2 Decentralized energy production

Centralized mechanisms are the typical ways of energy production that serve the original framework for the provision of energy services [7]. Such mechanisms rely on mass-energy production, bulk energy delivery through stretched transmission lines in the body of the nation-wide grid, and energy convey to energy masses of industrial, commercial, and residential consumers. Although the exploitation of centralized energy production mechanisms has underpinned unprecedented growth several decades ago, its decay is increasingly evident from the 21st century. This issue mostly back to the depletion of fossil fuel resources, upsurging resource conflicts tied to them, anthropogenic climate change, and ever-increasing economic and environmental concerns from their usage. All of the mentioned items offer ample reasons and a reasonable overview regarding the need for designing alternate energy production frameworks that allow the system for weaning off carbon-intensive fossil fuels in line with the objectives of modern energy networks. Herein, decentralized energy production systems are introduced as the first alternative solutions for supporting energy grids in achieving significant technical, economical, and environmental benefits in the energy generation, transmission, and distributions processes [8].

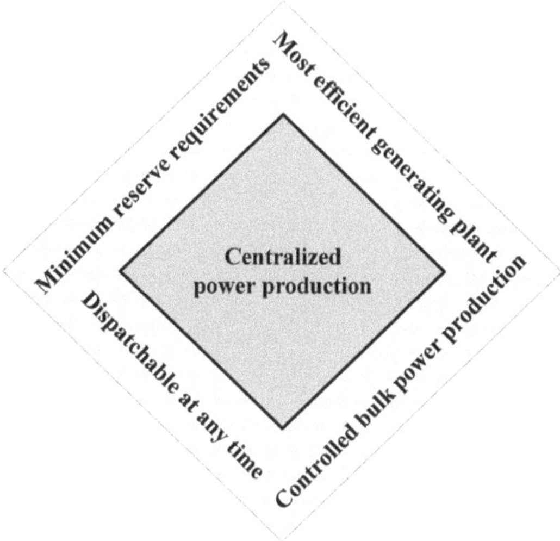

FIGURE 1.1 Key advantages of centralized energy production.

1.2.1 Dispersed energy generation

Energy networks have mostly relied on central energy production that the bulk power is produced and transmitted through high-voltage transmission lines to the distribution sector [9]. The produced power is then distributed to end-users via distribution transformers. This central mechanism for energy production is advantageous based on the three main aspects indicated in Fig. 1.1 [10].

As centralized mechanisms provide an acceptable condition in reliable and controllable energy serving, several drawbacks have risen critical challenges regarding their operation in modern energy networks. The prominent of these disadvantages back to releasing gaseous emissions, competition policies, deregulation, depletion of fossil-fuels resources, and techno-economic difficulties in their exploitation [10]. Such worriments have become the main motivation for transiting towards the connecting production into distribution networks that is referred to as dispersed generation [10]. Dispersed generation mostly stands on distributed energy production that offers remarkable benefits in comparison with traditional processes. It consists of fuel-based and carbon-free energy generation systems that its carbon-free or renewable types provide affordable and environmental-friendly conditions for energy production. Fig. 1.2 shows the explicit overview of differences between dispersed generation and centralized systems.

According to the schematic in Fig. 1.2, a distribution grid is a passive network that completely depends on the energy flow coming from the

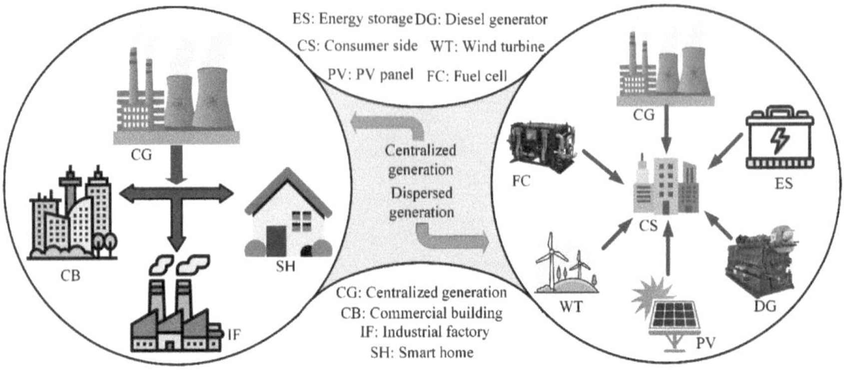

FIGURE 1.2 Overview of differences between dispersed generation and centralized systems.

high-voltage transmission lines. In such a process, the voltage drop occurs when the power flow is from the generation sector to the distribution level. However, as the generation units are distributed throughout the network in the dispersed generation, the system acts as the active network with bi-directional power flow and the voltage control relies on the distributed power production units as well as centralized ones.

1.2.1.1 *Different technologies*

In recent years, environmental damages caused by conventional systems have driven the research world to think about how substituting fossil fuel-driven systems can be possible with environmental-friendly technologies. This trend has been led to substantial developments in the renewable-type of dispersed-generator technologies. Widely geographically spreading dispersed generation units enables the system to reliably serve energy over the demand-side. The typical dispersed energy generation systems are indicated in Fig. 1.3.

The renewable dispersed generation units don't depend on external fuels and completely rely on climate factors such as solar radiation and wind speed in power production. Thus, they not only offer affordable energy generation but also promote clean energy production in the grid. The PV panels convert solar energy to DC power using solar cells and their power outputs directly depend on solar radiation, efficiency, and size of cells [11]. The produced DC power needs to be converted to the AC one as the distribution network is based on the AC power flow. This is done using power electronic interfaces such as DC-AC converters. Wind turbines are another way for clean energy production that are famous for their ability to generate power in the vast range. The wind speed in the related region is critical for producing power by wind turbines. On the other hand, this

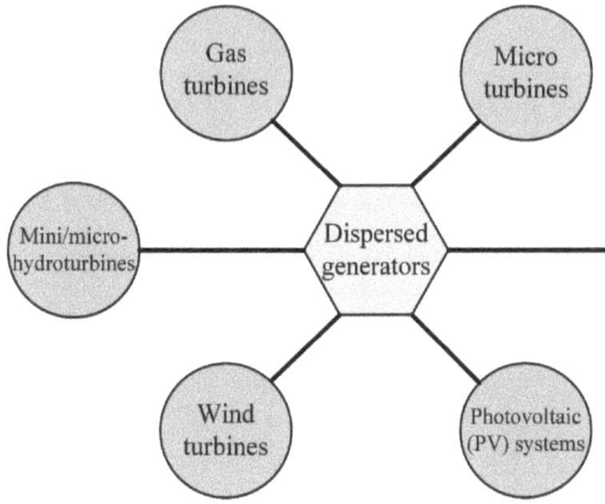

FIGURE 1.3 The typical dispersed-generator technologies.

issue also depends on the characteristics of the turbine such as rated power, cut-in, cut-out, and rated wind speeds [12]. If the wind speed is lower than cut-in speed or larger than cut-out speed, the turbine output will be zero aiming to prevent the turbine from damages at high speeds [13]. The wind power production will be a function of wind speed, cut-in, and rated speeds when the wind speed is between the cut-in and rated speeds. This is while the wind turbine output will be maximum and equal with the rated wind power when the wind speed is between the rated and cut-out speeds of the turbine.

Fuel cells are another renewable dispersed energy generation technology that work based on hydrogen energy. In renewable energy-rich times, the excess of energy can be used for generating hydrogen molar by the electrolyzer system. The hydrogen is then stored in the hydrogen storage to be used by the fuel cell for producing electricity or can be used by other energy conversion technologies in generating other carriers of energy [14]. Herein, hydropower units are the other strong contenders for producing clean energy that use specific turbines for power production based on the water flow. Hydropower units typically are in the range of MW for power generation and can be usefully used for controlling frequency in the power system [15].

In another side of dispersed energy generation, there are energy generation systems that are pertained to fossil fuels for producing power and cannot deliver clean energy to the system. Such units are called non-renewable dispersed energy generation technologies that can provide durable energy to the grid [16]. As their fuels are available at required times, their power production is controllable during the day. However, due to the environmental hazards of these units and decline in the fossil fuel resources, they are targeted to give their place in the energy generation process to renewable energy resources (RERs).

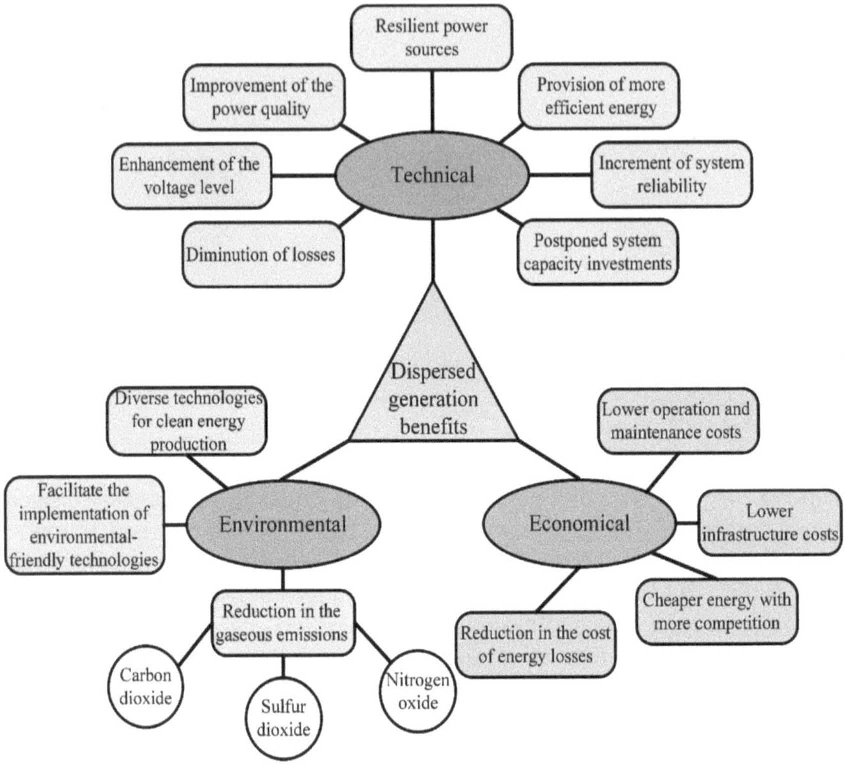

FIGURE 1.4 Dispersed generation benefits.

1.2.1.2 Benefits

The dispersed generation refers to the set of distributed energy generation technologies that promote decentralized energy production with the aim of overcoming the challenges of a centralized one. Indeed, the decentralized energy production paradigm offers two-way communication with pervasive control in the network structure while the centralized system is a hierarchical architecture with limited control on processes [17]. On the other hand, dispersed generation improves energy management by developing the self-monitoring capability instead of the previous blind system in the centralized structure and it provides a self-healing attribute for the modern energy grid and resolves the already manual restoration procedure in the traditional model. The energy production under dispersed mechanisms postpones the distribution and transmission upgrading and increases the system's reliability and quality of performance. To sum up, dispersed generation procures significant benefits, especially in terms of technical, economic, and environmental. Fig. 1.4 summarizes the prominence of these advantages in the modern energy grid.

According to Fig. 1.4, tremendous technical benefits can be reached in the operation of the system under dispersed energy generation technologies. Dispersed generation diminishes the power losses by bringing the production and consumption sectors closer to each other [18]. Indeed, in decentralized energy production, there is no need for long transmission lines for transmitting power from the generation sectors to the load centers and this issue can significantly reduce the power losses in the lines. Distributedly injecting power throughout the network in the dispersed generation enables the grid to improve power quality and effectively regulate the voltage level in all buses [19]. This mechanism can also increase the reliability of the system in energy supply due to not being dependent on only a few centralized power plants for energy production [20]. From the perspective of the economic aspect, reducing the power losses and subsequently their related costs can be led to the increment in the system's economic benefit. On the other hand, an ever-increasing trend in power consumption requires expansion schemes for the grid to effectively support load growth. These schemes need to cover all parts of the power system from generation to the distribution sectors with huge implementation costs. Dispersed generation can alleviate the grid expansion costs by distributedly energy serving in the local areas and deferring the requirement for implementing expansion plans [21].

In most cases, centralized energy production systems rely on fossil fuels for power generation. In this regard, not only the limited source of fossil fuels is available but also they can be the main cause for releasing a huge volume of harmful gases in the environment. Herein, employing eco-friendly technologies is an essential step for decreasing the unfavorable economical and environmental effects of fossil fuel-based systems. The renewable dispersed generation technologies make clean energy production possible in the potential areas of the grid [22]. Such strategies allow for producing carbon-free energy in a high range and decline the dependency of the system for conventional units.

1.2.1.3 Challenges

As mentioned, the future plans for developing energy grids are mostly relied on using renewable dispersed energy production systems aiming to avoid the unsatisfactory economic and environmental impacts of non-renewable based units in the system. Despite the beneficial role of RERs in modernizing future energy grids, especially in terms of environmental and economic, they bring a massive level of uncertainties for the energy generation sector creating a big challenge for optimal operation of the system [23]. Indeed, the intermittencies in RERs outputs reduce the confidence of the power availability and specify the great necessity for providing reliable remedies with the aim of reliable usage of RERs in the grid [24]. The

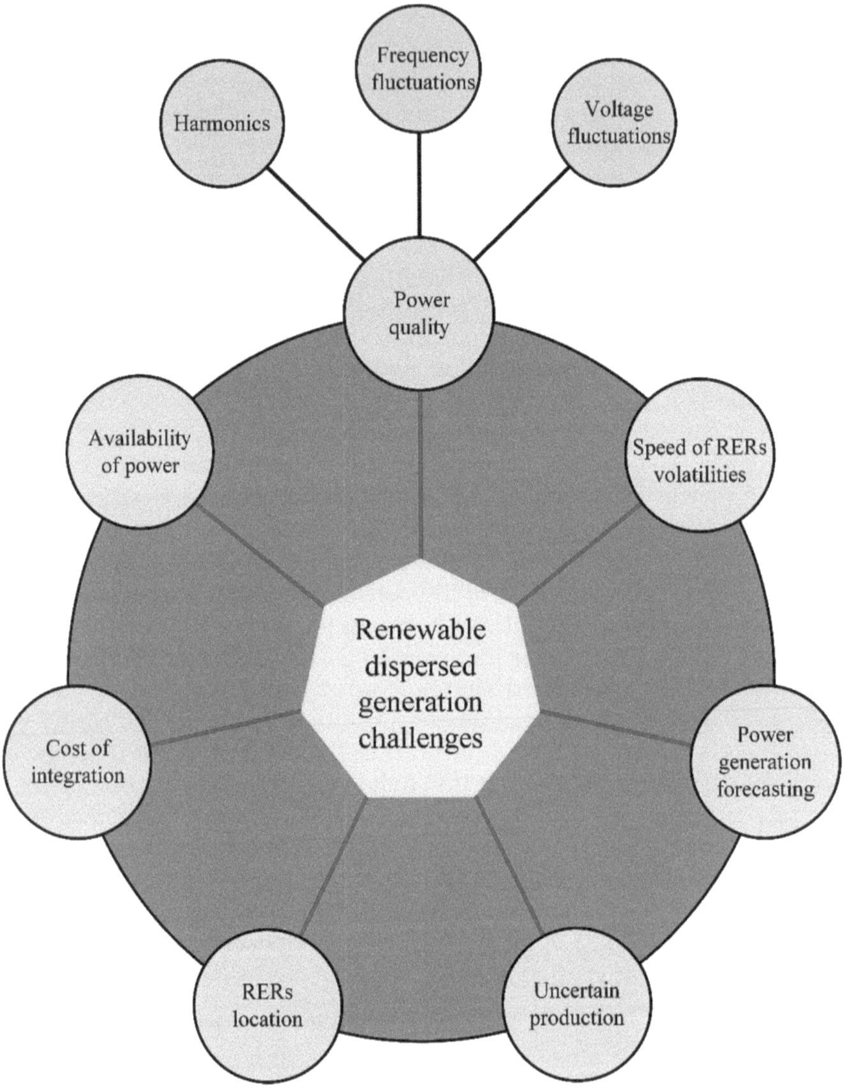

FIGURE 1.5 The prominent challenges of renewable dispersed generation.

prominent challenges of renewable dispersed generation are portrayed in Fig. 1.5.

- **Optimal integration and power quality challenges:** For reliably utilizing RERs in the body of the grid, different solutions are proposed that one of the important of them is the optimal integration of RERs in the energy structure. However, as this solution makes an appropriate avail of RERs possible, it exposes the system to new other critical challenges. For optimal integration of RERs, the grid structure needs to be developed based on the intelligent agents for realizing a holistic controlling and monitoring architecture aiming to effectively coordinate energy interactions for establishing an energy balance. Implementing such plans not only imposes a lot of costs on the system but also the technology limitations may hinder reaching the main goals of system monitoring. On the other hand, optimal integration of RERs can directly affect the power quality and face the system with bad power qualities. In this respect, given the ideal level of power quality as an imperative factor is essential for achieving high reliability and lower costs as well as avoiding deleterious effects for the grid, accurate analyzing its changes is vital in the renewable-penetrated systems. As one of the key characteristics of the power quality, the voltage level is exposed on the stochastic fluctuations in the presence of high RERs resulting in poor power quality for the network [25]. The adverse effects of RERs dependence on climate changes don't end just to the voltage but also the system frequency as the sensitive factor with narrow allowable variations can be affected by stochastic volatilities in the energy generation sector that drive the system for unbalances and instability conditions. Form another side, the output power of some RERs is DC and the power electronic interfaces will be required for connecting them to the main grid with AC power flow. This issue can also input different harmonics to the network making an inappropriate shape for the power quality.

- **Stochastic power production and its availability:** RERs are non-dispatchable units due to their uncertain nature in power production. Relying on climate change has inspired the uncontrollable attribute for them that require confident schemes for harnessing their inconsistent changes. The variability of their output is fully back to their dependence on the climate factors such as wind speed and solar irradiation. As such climate factors are uncertain in nature, the power production process of RERs will also be uncertain due to intermittent input. In such circumstances, the availability of the power directly back to the favorability of climate changes leading to inaccurate scheduling of the system.
- **Power production forecasting and protection challenges:** As mentioned, the optimal operation of the system with a full/high penetration of RERs is difficult and challenging work. Forecasting the occurrence of RERs outputs and other uncertain parameters can be one of the reasonable choices to meet the standard requirement for balancing energy between power supply and load [26]. However, the lack of capable approaches for conducting accurate forecasts restricts the design of reliable and sustainable plans for the power grid. These issues have also made the accurate predictions for long-term planning of the system almost impossible that is an imperative point in the development of

the future energy grid. From the perspective of the system protection, the dispersed generation mechanism converts the already passive grid with uni-directional power flow to the active and bi-directional system with multiple distributed energy resources (DERs). This issue adds difficulties and complexities for designing secure protection protocols for the distribution network. Hence, future modern energy networks are expected to benefit the advanced protection technologies for supporting bi-directional power flows in the sharp deregulated environment.

1.3 The trend for 100 percent RERs

In recent decades, electricity has been recognized as a key requirement for people's daily life. The popular applicability of electricity has made it a critical tool for developing different sectors of societies. This issue led to a rapid forming of primary grids and expanding frontiers in energy by feeding initial consumers from the network. Under the remarkable progresses in power generation, transmission, and distribution, electrification has been improved throughout the grid with the aim of provision of better services. In primary grids, the generation sector was absolutely centralized that the produced power was transmitted through transmission lines to the distribution sector. This mechanism stands on the uni-directional power flow in the system that cutting a certain area leads to the power outage in subsets of the distribution network. By the ever-growing of well-being indexes in people's lives, the disadvantages of primary energy grids and the related mechanisms were highlighted day by day. Due to the usage of fossil fuels in power production processes, the volume of released gaseous emissions in the atmosphere was dramatically increased, which has created concerns regarding the environmental implications [27]. The depletion of fossil fuels resources and their costs have also affected the raising of criticisms against the fossil fuels-based energy production units. Such drawbacks have driven the research world to find appropriate solutions for energy production instead of conventional systems. These solutions had enabled the network to deploy diverse types of energy production units and works based on the decentralized mechanism. Hence, current energy grids are a combination of centralized and decentralized systems in the energy production sector. The decentralized energy production mechanism brings valuable advantages for the power system, especially in terms of power quality, energy efficiency, sustainability, stability, reliability, and flexibility [28]. It allows the system to be more flexible in energy serving when the grid is exposed to different events. This ability of the decentralized-based mechanism is mostly back to the strategy of energy supply that relies on distributedly energy serving throughout the network. The carbon-intensive character of the non-renewable type of decentralized energy production has been driven the efforts for detecting new ways of power generation that offer environmentally-friendly processes [29]. Thus, another type of decentralized mechanism is appeared by availing environmental potential for energy production that is recognized as RERs. Although this type of decentralized energy generation is free of any harmful gaseous in its procedures unlike non-renewable ones, it suffers from controllable energy production due to its uncertain nature caused by stochastic changes in climate conditions [30]. On the other hand, the rapid growth in the technology of energy systems and the increasing trend of energy consumption by emerging new types of consumers are two key evolutions that are occurring simultaneously and alarms about the critical need for switching towards the sustainable clean network. Thus, future modern energy networks (FMENs) are planned to fully equip to RERs for completely generating clean energy and avoid any harmful gases in the energy production sector. As this target brings tremendous economic and environmental advantages for the grid, it poses basic concerns regarding the reliable operation of the system and makes the sustainable scheduling of the grid challenging. In other words, going towards the minimum dependency on fossil fuels-based energy production units as they are controllable alongside the maximum utilizing of RERs while they are uncertain, has suggested new challenges for the secure and reliable operation of energy grids. Therefore, given the trend of FMENs towards 100 percent RERs, they require to deploy sustainable paradigms for handling the usage of fully RERs in the energy generation process. In this regard, besides the exploitation of a variety of energy storage technologies and diverse approaches for the optimal integration of RERs, another effective way for facilitating the usage of 100 percent RERs in FMENs is to employ different energy trading strategies, which is one of the main goals of this book. The development trend for energy grids is indicated in Fig. 1.6.

1.4 Modern energy grids

The idea of modernizing energy grids is born when the lack of strict principles in energy production threatened the eco-friendly and cost-effective operation of energy networks. However, it is not only limited to the economic and environmental aspects of the grid exploitation, but also it implies the motivations for enhancing the convenience of people's lives in line with the developments in various technologies. As thinking about the "Modern" word may lead to flourishing different imageries in our minds, it is essential to clarify what does it mean in the energy world and it covers which criteria in the energy studies. To effectively concentrate on addressing ambiguities in the context of modern energy grids, the US Department of Energy's (DOE) Grid Modernization Laboratory

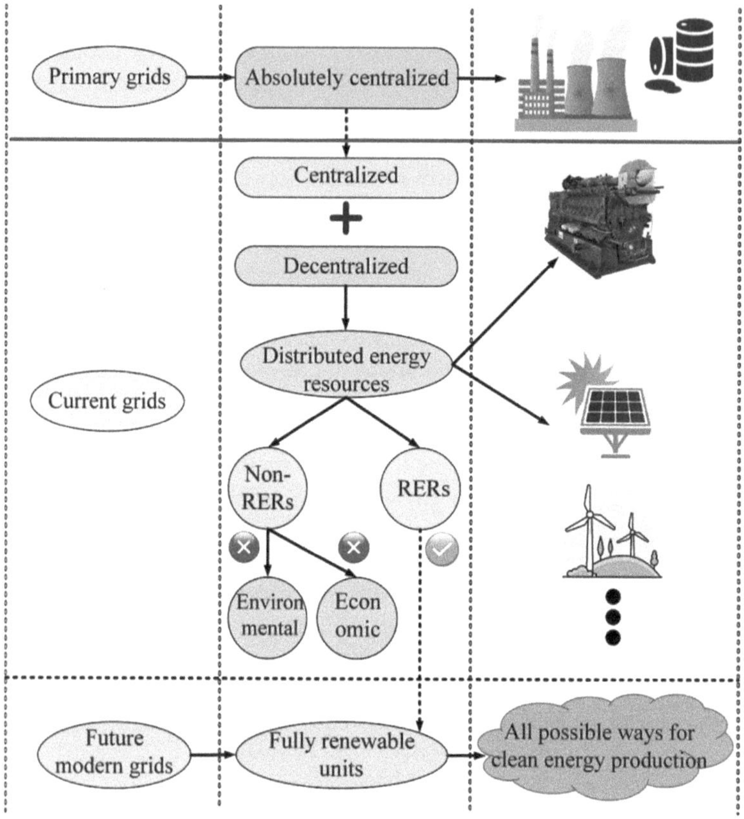

FIGURE 1.6 The development trend for energy grids.

Consortium pursued research on the distribution system planning and the role of utility regulators in its process to investigate it considering the grid modernization principles [31]. Moreover, ambitious endeavors are driven to extend the boundaries of social science and energy technologies toward grid modernization and offer feasible solutions for the challenges ahead.

1.4.1 Definition

In order to effectively scrutinize a special topic, it is required to be familiar with its definition and which areas are aimed to be supported by the related topic. Regarding modern energy networks and grid modernization, different definitions and also unique descriptions are not available due to their initial formation in the research world. However, recently, the California Public Utilities Commission (CPUC) has provided a definition for grid modernization that relatively describes its features. This definition is "A modern grid allows for the integration of DERs while maintaining and improving safety and reliability. A modern grid facilitates the efficient integration of DERs into all stages of electric power system planning and operations to fully utilize the capabilities that the resources offer, without undue cost or delay, allowing markets and customers to more fully realize the value of the resources, to the extent cost-effective to ratepayers, while ensuring equitable access to the benefits of DERs. A modern grid achieves safety and reliability of the grid through technology innovation to the extent that it is cost-effective to ratepayers relative to other legacy investments of a less modern character" [31].

As the mentioned definition suffers from the lack of plenary coverage regarding multi-carrier energy networks, technology resource classifications, and methodologies for the widespread exploitation of intelligent hybrid units, the new definition is proposed in [32] to make the grid modernization descriptions completely than ever before. This new definition for modern energy networks framework is "an integrated infrastructure of control, economic, and environmental mechanisms that allows the hybrid system to easily adopt multidisciplinary innovative technologies for maintaining system reliability and stability, improving system efficiency and safety, increasing system flexibility and privacy, enhancing the grid interoperability, providing grid edge services, adopting various types of clean energy production units, facilitating for effective interactions, and enabling the system for possible developments across the entire FMENs' infrastructure" [32]. This is while the science of grid modernization is developing day by day and the emergence of new and complete definitions is expected in future works.

1.4.2 Objectives

As the pioneering book, this book strives to clarify the objectives of grid modernization and states which goals are targeted to be realized for FMENs. These objectives are further derived from disparate aspects of grid modernization that cover the technical, environmental, economic, and social criteria in the designed plans. For this aim, this chapter summarizes the set of objectives for FMENs that are listed as follows. Meanwhile, the objectives for grid modernization are updated given the well-being and convenience factors expected for modern society.

- Improving the security of the system from energy generation to the distribution considering privacy-preserving of all participants and assets of other parts of the grid.
- Designing based on the self-healing ability by focusing on implementing automated maintenance and operation paradigms.

- Holistically monitoring system for moment-to-moment processing of the system's status with the capability of covering multi-vector energy units and new smart devices.
- Designing flexible and sustainable architecture suitable for adopting diverse types of RERs.
- Developing expert energy managing and controlling technologies in line with the scalability of the system.
- Establishing an extensive environment for multi-energy interactions with multifarious customer choices as well as sufficient and reasonable freedom for different participants of the grid.
- Manufacturing adaptable architecture for handling emerging technologies as well as food-energy-water nexus cycles.
- Developing an integrated system that is adaptable, controllable, sustainable, flexible, reliable, and observable at scale.
- Supplying clean and affordable multi-energy to end-users in a reliable manner.
- Utilizing information, sensing, and measurement technologies for an optimally operating reconfigurable system.

1.4.3 Principles

Generally, the successful execution of each plan is directly subject to determining appropriate principles for it considering the effective factors in the related process. The defined principles need to be complete in their coverage aiming to support all aspects of designing, scheduling, and operation of the system. This is because the defined principles can be general guidance for participants and effectively drive them to incorporate in energy interactions. The set of suggested principles for FMENs is listed as follows but they are not limited to them given the progress trend of modern energy grids.

- The proposed models for FMENs should be able to support the integrated system to be reliable, observable, extensible, flexible, adaptable, auditable, sustainable, resilient, scalable, reconfigurable, and efficient with the aim of facilitating the implementation of modern energy grids in accordance with their goals.
- FMENs should be sufficient capable in maintaining and improving the system safety, stability, efficiency, and reliability of continuous energy supply in the presence of a high level of uncertainties.
- FMENs need to be an exhaustive environment for all energy customers, consumers, producers, and shareholders to allow them to effectively perform their role in the interoperability cycle of the system.
- FMENs should serve uninterrupted, eco-friendly, and cost-effective multi-energy to consumers without relying on load shedding schemes.

- FMENs need to adopt innovative technologies for making fully clean energy generation possible by employing sustainable strategies to effectively manage uncertainties in balancing energy of the system.
- FMENs need to coordinately integrate different energy networks for the optimal operation of the interconnected system with sufficient flexibility, reliability, and sustainability.
- FMENs should be able to promote effective synergies among hybrid energy elements as well as procure efficient interoperability in the system.
- FMENs need to improve the convenience factors, electrification system, and end-user services in line with the technology development.

1.4.4 Opportunities and challenges

Modern energy grids are introduced as a holistic structure for the energy world to convert most of the key items that already were a challenge to opportunities for significantly upgrading the system in the whole area [33]. In this transition, substantial developments in information and communication technologies, as well as multi-energy systems, can support the system in realizing the goals of modern energy grids. Indeed, FMENs are scheduled to benefit the automatic system for monitoring energy interactions with the aim of making the energy structure enough reliable and resilient. They are targeted to build a holistic environment that enables participants for interoperability while provides technologically-induced customer services. On the other hand, FMENs is planned to use fully RERs potential for clean energy production and adopt innovative strategies for establishing sustainability in the whole infrastructure of the system. Fig. 1.7 illustrates the opportunities of modern energy grids.

In the pass of realizing the mentioned opportunities, the system faces some critical challenges making the goals of modern grids hard to achieve. The challenges regarding technology and assets adequacy, capable strategies for welcoming fully renewable systems, and complexity of the structure in terms of managing, controlling, and monitoring in the presence of numerous intelligent devices are some of the prominent of these challenges that are ahead of FMENs.

- A holistic energy structure that provides sufficient services, can rapidly foster, efficiently recover disruptions, can effectively realize FMENs requirements.
- The complexity in simultaneously and optimally monitoring several energy networks, such as electrical, natural gas, district heating, and water grids with control centers, smart systems, and remarkable constraints.
- Effectively handling the computational burden of the expert units, the complexity of the system monitoring, and various computational languages for scrutinizing the grid.

FIGURE 1.7 The opportunities of modern energy grids.

- Optimally integrating energy grids consisting of consumers with diverse consumption patterns, energy market structure owners with multifarious expectations and opinions, governments with different policies, energy interactions players with a variety of styles, and regions with several geographical conditions.
- The reasonable ways for persuading stakeholders, producers, and consumers with various requirements to accept FMENs conditions and act in accordance with FMENs guidelines when they are involved in energy management schemes, energy interactions, and smartization programs.
- Developing an effective energy architecture and a holistic platform aiming to adopt emerging hybrid systems and technologies with a variety of applications and diverse requirements.
- Managing the great transition from conventional systems to modern energy infrastructure with a huge evolution in the marketing, operational, and planning of the system as well as adopting informed institutional decision-making processes that are prudent and future-looking.

1.5 Summary

This chapter strives to give an appropriate overview of grid modernization and the related steps in the pass of realizing modern energy grids.

In light of multi-energy growth in the grid alongside the tremendous progress in technology development, the role of different technologies in the transition towards sustainable and modern energy networks was discussed. Due to the importance of decentralized energy production in the system, the applicability of dispersed generation especially its renewable types were scrutinized along with assessing their benefits and challenges. Why FMENs are scheduled for 100 percent RERs was described along with highlighting the related trend for modern energy systems. Moreover, to effectively investigate the pass for grid modernization, the related definitions, principles, objectives, opportunities, and challenges were procured with detailed assessments.

1.6 Questions

The careful analyzing whole of this chapter is expected to enable readers to effectively evaluate the following questions.

1) What is the role of dispersed generation technologies in the decentralization of FMENs?
2) How can the different technologies of the dispersed generation support the grid for sustainability and reliability in energy supply?
3) Which benefits can be achieved from the operation of dispersed generation and these benefits are for the system from which aspects?
4) What are the expected challenges for FMENs when they are exploited under dispersed energy generation technologies?
5) Which drawbacks of the centralized energy generation have driven the system to move towards decentralized energy production strategies?
6) Why FMENs are targeted at going towards 100 percent RERs in their energy production sector?
7) Which challenges threaten the realization of fully clean energy generation in modern energy grids?
8) What would be the system architecture for unbroken multi-energy serving in the presence of 100 percent RERs?
9) What are the effective factors in the call for modern energy grids and why these factors are prominent in controlling the system?
10) What are the definitions for modern energy grids and they cover which aspects of FMENs?
11) What are the objectives of building modern energy grids and why they are important to be achieved?
12) Under which principles FMENs are expected to be operated and what are their effects in realizing the goals of FMENs?
13) FMENs are aimed to bring which opportunities for the grid participants?

14) Why optimal energy interactions and establishing interoperability between energy market players are critical for the flexibility of FMENs?
15) What are the challenges ahead of grid modernization and how they can be addressed in the hybrid system?

1.7 Suggestions

Given the topic of this chapter, the following suggestions are provided for the readers assisting them to pursue related research clues.

1) Developing novel decentralized energy generation technologies can support FMENs in improving their sustainability, reliability, and flexibility.
2) Proposing innovative solutions for effectively tackling the unpredictability of renewable dispersed energy generation can facilitate the clean energy production process in the system.
3) Addressing the challenges of dispersed energy generation technologies can support the system in reaching the decentralization goal.
4) Developing innovative strategies for making 100 percent carbon-free energy generation possible can facilitate the implementation of grid modernization schemes.
5) Investigating principles, objectives, opportunities, and challenges of modern energy grids can effectively highlight the need for grid modernization in near future.

1.8 Future trends

Future trends can give a proper overview for the readers to form their related works and develop effective solutions for remaining gaps, which are listed as follows for this chapter.

1) The new clean dispersed energy generation technologies need to be developed for procuring the possibility of fully eco-friendly energy production in the system.
2) The innovative strategies are essential for properly dealing with uncertain outputs of RERs to improve the reliability and sustainability of the energy grid.
3) The practical mechanisms are necessary for optimally integrating different energy networks aiming to enhance the flexibility of the system in multi-energy serving.
4) The novel architecture needs to be designed for FMENs to allow the grid players for interoperability at the widespread level.
5) Smart platforms are critical for optimally monitoring the complex system with numerous intelligent devices.

References

[1] D. Niu, M. Yu, L. Sun, T. Gao, K. Wang, Short-term multi-energy load forecasting for integrated energy systems based on CNN-BiGRU optimized by attention mechanism, Appl. Energy 313 (2022) 118801.

[2] M. Daneshvar, E. Babaei, Exchange market algorithm for multiple DG placement and sizing in a radial distribution system, J. Energy Manage. Technol. 2 (1) (2018) 54–65.

[3] J. Yan, O.A. Broesicke, X. Tong, D. Wang, D. Li, J.C. Crittenden, Multidisciplinary design optimization of distributed energy generation systems: the trade-offs between life cycle environmental and economic impacts, Appl. Energy 284 (2021) 116197.

[4] M. Daneshvar, B. Mohammadi-ivatloo, S. Asadi, S. Galvani, Short term optimal hydro-thermal scheduling of the transmission system equipped with pumped storage in the competitive environment, J. Electr. Eng. 14 (1) (2020) 77–84.

[5] P. Sanongboon, T. Pettigrew, Hybrid energy system optimization model: electrification of Ontario's residential space and water heating case study, Energy Clim. Change 3 (2022) 100070.

[6] M. Daneshvar, B. Mohammadi-Ivatloo, K. Zare, M. Abapour, S. Asadi, A. Anvari-Moghaddam, Chance-constrained scheduling of hybrid microgrids under transactive energy control, Int. J. Energy Res. 45 (7) (2021) 10173–10190.

[7] A.M. Adil, Y. Ko, Socio-technical evolution of Decentralized Energy Systems: a critical review and implications for urban planning and policy, Renew. Sustain. Energy Rev. 57 (2016) 1025–1037.

[8] J. Yan, M.A. Brown, D. Yu, J.C. Crittenden, Policy incentives and social cost of emissions for promoting decentralized energy production: a life cycle cost analysis, J. Clean. Prod 282 (2021) 125394.

[9] J.R. Kotzebue, M. Weissenbacher, The EU's Clean Energy strategy for islands: a policy perspective on Malta's spatial governance in energy transition, Energy Policy 139 (2020) 111361.

[10] N. Gupta, N. Kaur, S.K. Jain, K.S. Joshal, Smart grid power system, in: Advances in Smart Grid Power System, Elsevier, 2021, pp. 47–71.

[11] M. Daneshvar, H. Eskandari, A.B. Sirous, R. Esmaeilzadeh, A novel techno-economic risk-averse strategy for optimal scheduling of renewable-based industrial microgrid, Sustain. Cities Soc. 70 (2021) 102879.

[12] N. Karakasis, E. Tsioumas, N. Jabbour, A.M. Bazzi, C. Mademlis, Optimal efficiency control in a wind system with doubly fed induction generator, IEEE Trans. Power Electron. 34 (1) (2018) 356–368.

[13] M. Daneshvar, B. Mohammadi-ivatloo, K. Zare, Integration of distributed energy resources under the transactive energy structure in the future smart distribution networks, in: Operation of Distributed Energy Resources in Smart Distribution Networks, Elsevier, 2018, pp. 349–379.

[14] Y. Liu, et al., A Novel Adaptive Model Predictive Control for Proton Exchange Membrane Fuel Cell in DC Microgrids, IEEE Trans. Smart Grid (2022).

[15] D. Liu, C. Li, O. Malik, Operational characteristics and parameter sensitivity analysis of hydropower unit damping under ultra-low frequency oscillations, Int. J. Electr. Power Energy Syst. 136 (2022) 107689.

[16] M.M. Islam, M. Irfan, M. Shahbaz, XV. Vo, Renewable and non-renewable energy consumption in Bangladesh: the relative influencing profiles of economic factors, urbanization, physical infrastructure and institutional quality, Renew. Energy 184 (2022) 1130–1149.

[17] Y. Wu, Y. Wu, J.M. Guerrero, J.C. Vasquez, Digitalization and decentralization driving transactive energy Internet: key technologies and infrastructures, Int. J. Electr. Power Energy Syst. 126 (2021) 106593.

[18] M. Delfanti, D. Falabretti, M. Merlo, Dispersed generation impact on distribution network losses, Electr. Power Syst. Res. 97 (2013) 10–18.
[19] C. Gaunt, R. Herman, E. Namanya, J. Chihota, Voltage modelling of LV feeders with dispersed generation: probabilistic analytical approach using Beta PDF, Electr. Power Syst. Res. 143 (2017) 25–31.
[20] M. Daneshvar, M. Abapour, B. Mohammadi-ivatloo, S. Asadi, Impact of optimal DG placement and sizing on power reliability and voltage profile of radial distribution networks, Majlesi J. Electr. Eng. 13 (2) (2019) 91–102.
[21] M. Ahmadi, O.B. Adewuyi, M.S.S. Danish, P. Mandal, A. Yona, T. Senjyu, Optimum coordination of centralized and distributed renewable power generation incorporating battery storage system into the electric distribution network, Int. J. Electr. Power Energy Syst. 125 (2021) 106458.
[22] M. Davoudi, V. Cecchi, J.R. Agüero, Effects of stiffness factor on bus voltage variations in the presence of intermittent distributed generation, in: 2015 North American Power Symposium (NAPS), IEEE, 2015, pp. 1–6.
[23] S. Cai, Y. Xie, Q. Wu, M. Zhang, X. Jin, Z. Xiang, Active and Reactive Power Coordinated Two-Stage MG Scheduling for Resilient Distribution Systems under Uncertainties, IEEE Trans. Smart Grid (2022).
[24] M. Daneshvar, B. Mohammadi-Ivatloo, K. Zare, A. Anvari-Moghaddam, Transactive Energy Strategy for Energy Trading of Proactive Distribution Company with Renewable Systems: a Robust/Stochastic Hybrid Technique, e-Prime-Advances in Electrical Engineering, Electronics and Energy (2022) 100028.
[25] R. Smolenski, P. Szczesniak, W. Drozdz, L. Kasperski, Advanced metering infrastructure and energy storage for location and mitigation of power quality disturbances in the utility grid with high penetration of renewables, Renew. Sustain. Energy Rev. 157 (2022) 111988.
[26] Y. Yang, S. Bremner, C. Menictas, M. Kay, Forecasting error processing techniques and frequency domain decomposition for forecasting error compensation and renewable energy firming in hybrid systems, Appl. Energy 313 (2022) 118748.
[27] F. Moellenbruck, T. Kempken, M. Dierks, G. Oeljeklaus, K. Goerner, Cogeneration of power and methanol based on a conventional power plant in Germany, J. Energy Storage 19 (2018) 393–401.
[28] M. Moner-Girona, et al., A multidimensional high-resolution assessment approach to boost decentralised energy investments in Sub-Saharan Africa, Renew. Sustain. Energy Rev. 148 (2021) 111282.
[29] S. Shan, M. Ahmad, Z. Tan, T.S. Adebayo, R.Y.M. Li, D. Kirikkaleli, The role of energy prices and non-linear fiscal decentralization in limiting carbon emissions: tracking environmental sustainability, Energy 234 (2021) 121243.
[30] M. Daneshvar, B. Mohammadi-Ivatloo, K. Zare, A novel transactive energy trading model for modernizing energy hubs in the coupled heat and electricity network, J. Clean. Prod (2022) 131024.
[31] A.L. Cooke, J.S. Homer, L. Schwartz, Distribution System Planning-State Examples by Topic, Pacific Northwest National Lab.(PNNL), Richland, WA United States, 2018.
[32] M. Daneshvar, S. Asadi, B. Mohammadi-Ivatloo, Overview of the Grid Modernization and Smart Grids, in: Grid Modernization- Future Energy Network Infrastructure, Springer, 2021, pp. 1–31.
[33] J.R. Agüero, E. Takayesu, D. Novosel, R. Masiello, Grid modernization: challenges and opportunities, Electr. J. 30 (4) (2017) 1–6.

CHAPTER

2

Analysis of basic requirements for switching from current energy grids to future modern multi-carrier energy networks

2.1 Introduction

Over recent decades, energy grids have witnessed a tremendous transformation from the residential sector to the network-level part. This evolution was accompanied by substantial changes in energy grid faces that were affected by the emergence of new smart devices as well as the proliferation of customer-owned units calling for innovative ways for system operation, control, and protection [1]. On the other hand, this evolution has been intensified when energy become a crucial tool in people's daily life. It surrounded the different parts of the system from the residential sector to the network level. In the residential sector, the number of energy consumers is upsurging day by day, and this further back to expanding of energy grids and developing energy-consumed systems throughout the grid [2]. The remarkable improvements in electrification were posed energy supply to consumers over long distances as this goal was later satisfied by using distributed energy resources (DERs) [3]. The revolution for the residential sector has now arrived at the point of smartization of the energy-customer sector. Herein, the call for modern society is more sensed after the appearance of smart apparatus and growing the convenience

Emerging Transactive Energy Technology for Future Modern Energy Networks
DOI: https://doi.org/10.1016/B978-0-323-91133-7.00008-9

factors of residents in different nations. However, the successful transition from the previous residential structure to the modern one demands new requirements that need to be considered in the grid modernization pass. The industrial transformation of energy networks started from first industry revolution that was accompanied by the invention of a steam engine as well as enhancing the usage of water and steam power [4]. Given the primary developments in the steam engine and power production process in the related period, the initial structure of the power system was built based on centralized energy production [5]. This type of energy production mechanism allows the system to generate energy centrally using electric generators and transmit it to the load centers. However, the growth of industrial civilization did not end at this stage as the rapid advancements in global technologies drove the system towards conditions with better services. Hence, the industrialization trend witnessed second and third revolutions that were relied on the developments for changing the style of energy generation from exclusively affinity to centralized fossil fuel-based generation towards less dependency on them by facilitating the operation of natural resources and environmental potential for energy production. However, we are in the fourth industrial revolution that the challenges of all sectors of energy grids are no longer acceptable due to technology advancements in intelligent systems design and automatic units operation. Meanwhile, the provision of innovative solutions is an inevitable step in meeting requirements and realizing such expectations for modern energy grids. From large-scale viewpoints, revolutions in energy networks are not limited only to the residential and industrial sectors and the grid has not been unaffected by this development. The technology developments have led to the building of new devices that can work with different carriers of energy and provide a lot of flexibility for the system [6]. On the other hand, co- and tri-generation energy systems are enough developed to come to the cycle of energy production and make simultaneously multi-energy generation possible for the system [7]. Such evolutions have further highlighted the undeniable dependencies between various energy grids. This is while eco-friendly plans have also made a high/full clean energy production necessary for future schemes [8]. Thus, optimal operation of modern energy grids will be subject to considering the restrictions of different energy grids such as electricity, natural gas, water, and district heating networks and energy interactions among them. How the technology will allow for realizing such integrated energy grids is the key question that is the basis of most research clues in the grid modernization pass. This chapter is aimed to effectively analyze the basic requirements for switching from current energy grids to future modern multi-carrier energy networks by concentrating on the evolution trend in the residential, industrial, and grid-level sectors.

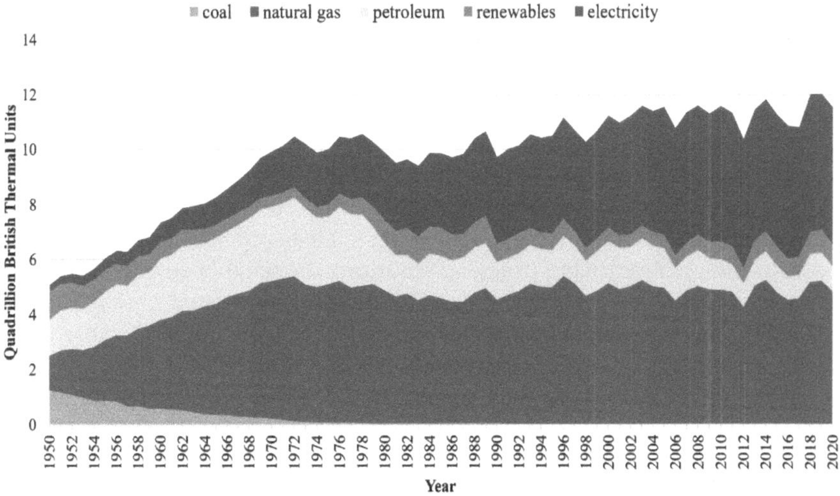

Source: U.S. Energy Information Administration, Monthly Energy Review, Table 2.2, May 2021.
Note: Electricity excludes losses in electricity generation and delivery. Petroleum includes heating oil, liquefied petroleum gas (propane), and kerosene. Renewables include wood, geothermal energy, and solar energy.

FIGURE 2.1 U.S. residential sector energy consumption by different energy sources from 1950 to 2020 [10].

2.2 Residential developments

Building energy consumption constitutes more than 30 percent of the whole expected energy consumption worldwide making this sector responsible for releasing a third of the world's gaseous emissions [9]. This statistic is more than 40 percent of the total energy for buildings in the United States that is led to almost 38 percent contribution for the gaseous emissions. In addition, more than 75 percent of produced power in the United States is consumed in buildings [9]. From the prevalence of energy and maturing its technologies, the growth rate for energy consumption has always been upward for all sectors, especially the residential part. Fig. 2.1 portrays the energy consumption changes for the residential sector of the U.S. based on the different energy sources from 1950 to 2020 [10].

According to Fig. 2.1, the amount of coal consumption had a downward trend from 1950 to 2007 so that its consumption has become zero in 2008. This trend is formed for coal due to avoiding the harmful effects of its burning on the environment. However, it can be deduced from Fig. 2.1 that electricity has always been a popular energy carrier as its consumption changes have permanently experienced a growing trend.

The increasing trend of energy consumption can be because of the popularity of energy in enhancing the well-being of people and easies of performing different works in the short time that made it a vital tool among nations [11]. From another perspective, the number of electricity-based appliances that already were not electrified is quickly risen due to advances in their technologies. Increasing such appliances has added new loads to the grid requiring practical solutions for reliably supplying them.

On the other hand, technology development has opened new doors for intelligent appliances that have the capability of working with more than one energy carrier. This movement is done not only for improving the flexibility of the grid in energy serving in the presence of renewables but also is carried out for minimizing the rate of inconveniencies by switching to use another energy carrier when the first one is cut out from the grid. Hence, future modern energy networks (FMENs) need to adopt innovative strategies and capable systems for satisfying the goals of modern energy networks in the residential sector. Some of the prominent requirements for modernization of the residential sector are indicated in Fig. 2.2.

2.3 Industry revolution

The industrial transformation is started from the first Industrial Revolution when the steam engine was invented for the first time and boosted the energy production process under steam power. This revolution was in the 18th century when water and steam power were common ways of energy generation based on mechanical processes [4]. The man-to-man oral communication was the typical way for managing procedures and the system suffered from the lack of adequate technology for automatically controlling and monitoring the system. The technological advancements allowed for establishing man-to-man writing communication in the second Industrial Revolution. This was the first time for electrical energy appearance with mass production. However, the system was faced with critical challenges such as ecosystem degradation, environmental pollution, and resource constraints that created motivations for future upgrades. Hence, the idea of using the environmental potential for energy production was born in the third Industrial Revolution that reduced greenhouse gas emissions by declining the burden of energy production from the fossil fuel-based centralized systems [4]. This period of revolution was also accompanied by automated processes with digital technologies that offered man-to-machine communication. As the achievements were impressive in the third revolution, the technology developments in the fourth revolution drove the system to benefit the internet of energy for smartization of the grid,

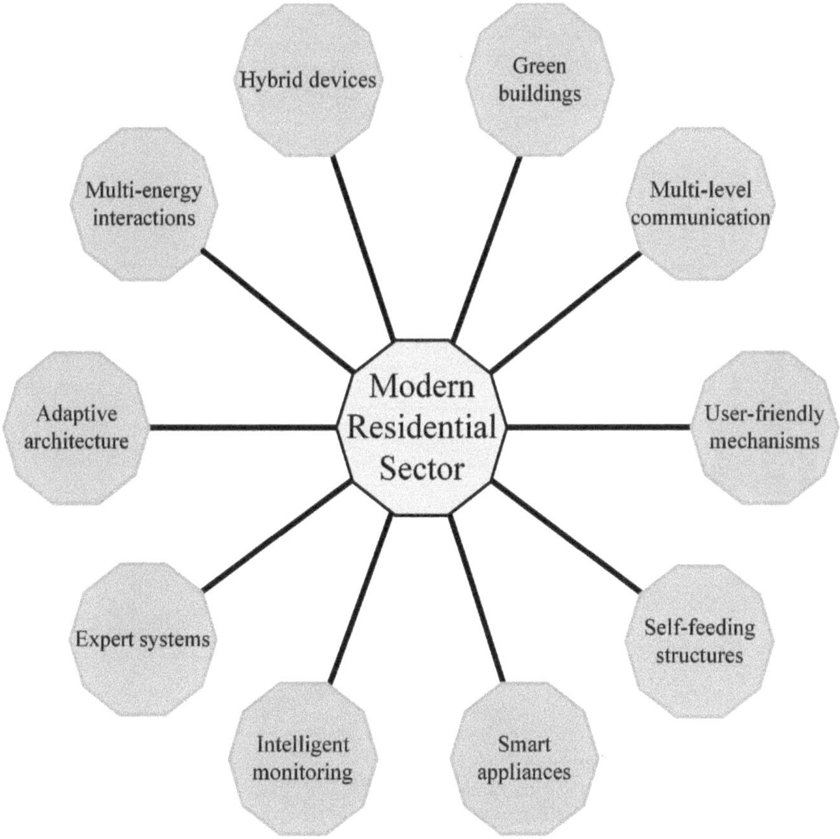

FIGURE 2.2 Some of the prominent requirements for modernization of the residential sector.

which utilized several types of DERs for energy production. The cyber-physical system is appeared and developed based on artificial intelligence (AI) for increasing the resiliency and sustainability of the system. The information and communication technologies are also enough developed to support the smart energy interactions throughout the automatic structure. Fig. 2.3 demonstrates the details of the developments for the four industrial revolutions [4].

In addition to the significant developments in the industry part of energy networks in previous revolutions, new challenges are transpired in front of attempts for realizing the valuable advantages. One of these benefits is the design of architecture enabling the system for fully clean energy production that is targeted aiming to conserve the environment from harmful gaseous. However, realizing this goal needs to procure sustainable and reliable strategies that ensure continuous power serving in the system

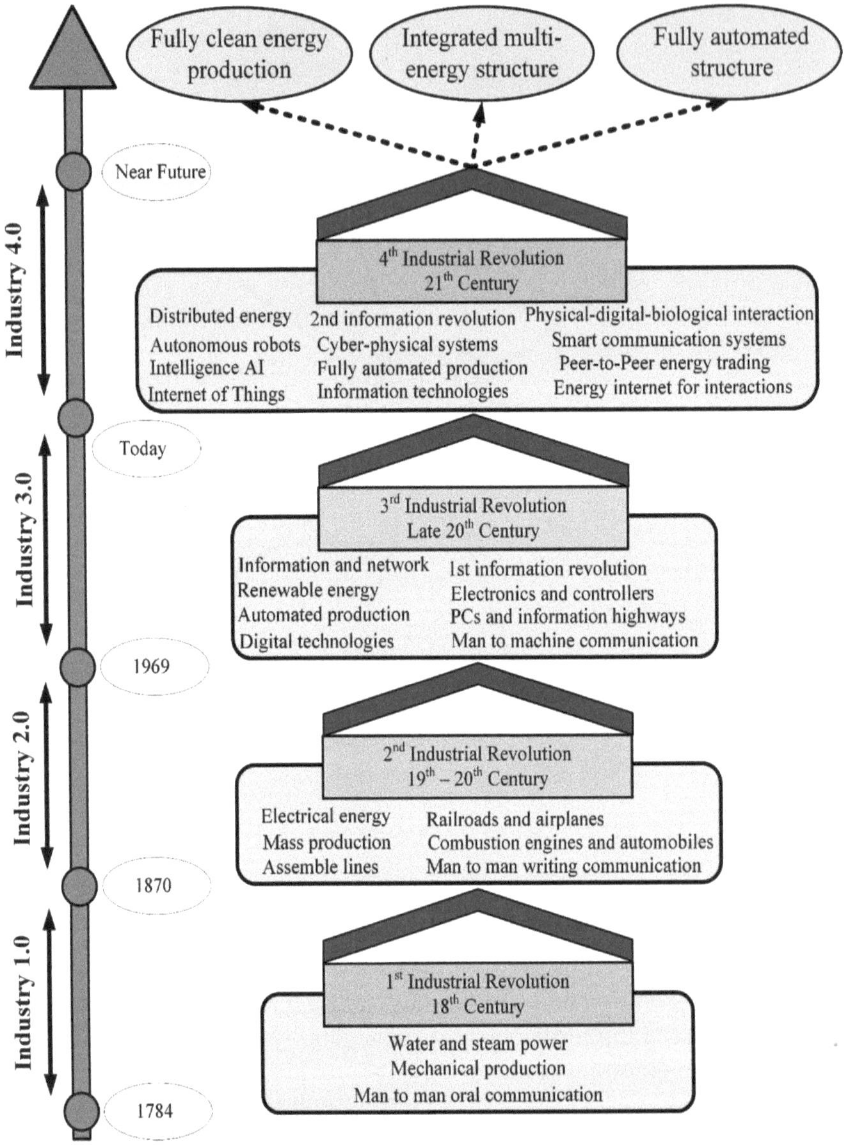

FIGURE 2.3 Features of developments for the four industrial revolutions.

with 100 percent renewable energy sources (RESs). Indeed, the uncertainties in RESs outputs need to be effectively captured for developing holistic models to create adequate assurance from the sustainability, reliability, and resiliency of the system in the presence of 100 percent RESs.

Another prominent goal is to interconnect different energy networks, consider their interactions, and improve them using energy conversion, storage, and trading technologies with the aim of providing sufficient flexibility for the system [12]. However, how different energy networks with diverse limitations can be interconnectedly operated in an optimal manner is a key challenge that requires confident treatments for solving it. Furthermore, energy efficiency is also a critical factor in the coupled structure calling for high-efficiency devices and processes. On the other aspect, deploying a fully automated energy structure is another vital requirement for FMENs to allow for easy monitoring and controlling of the system in multifarious conditions of the integrated grid. All of the aforementioned goals are some of the prominent objectives for FMENs, which ambitious endeavors are required for facilitating their realization in the modern energy infrastructure.

2.4 Network expansion

Another key part of the system affected by the grid modernization schemes is the network-level changes. Indeed, modernization envisions a new evolution stage for the energy network that affects the different parts of the grid with substantial changes. These changes actually belong to three main parts: energy grid for energy exchanging, information grid for service serving, and communication grid for data trading [4]. In the energy grid part, modernization schemes seek to achieve optimal generation by effectively managing available energy resources, flexible distribution by employing innovative energy sharing strategies, energy management programs, energy conversion technologies, and energy storage systems. It also seeks to provide smart transmission by using intelligent devices, effective transformation, and efficient energy consumption by deploying high-efficiency appliances. In the communication network, modernization plans pursue to realize the comprehensive perception, controllable actuation under multi-agent systems, reliable identification using AI approaches, inter-operative communication by applying expert coordinating devices, and ubiquitous collection. This is while, in the information network, its schemes are targeted to develop a holistic smart integration platform that is famous in its flexibility and heterogeneity, scalability, and openness and includes leading-edge technologies such as big data of things, AI of things, and the cloud of things, autonomous market-based schemes, sustainable business-based projects, and different types of innovative services. The detailed elements of these three parts i.e. energy-oriented network, service-oriented management, and communication-oriented network are illustrated in Fig. 2.4. This Fig. is the developed version of the related one in [4].

2.4.1 Multi-carrier energy networks

In this regard, the growing demand for different carriers of energy as well as emerging multi-energy systems have indicated the need for

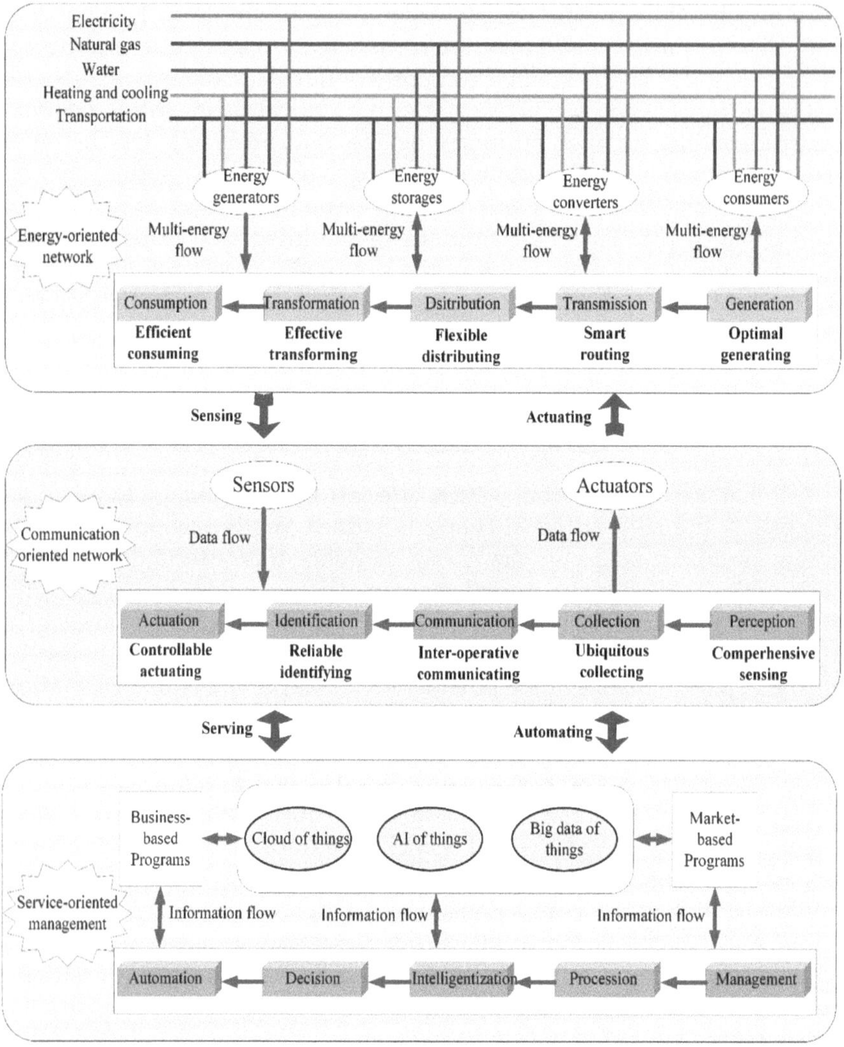

FIGURE 2.4 Three network parts in the grid modernization process.

innovative plans for upgrading energy systems for the unbroken multi-energy supply [13]. The quick increment in the operation of co- and tri-generation units for simultaneously generating different carriers of energy has affected various energy networks at their operation time [14]. This is because these systems such as combined, heat and power (CHP) use for example natural gas for generating heat and power so injecting heat and power by this unit to heat and power grids affect the optimal operation of the gas network due to consuming natural gas for energy production.

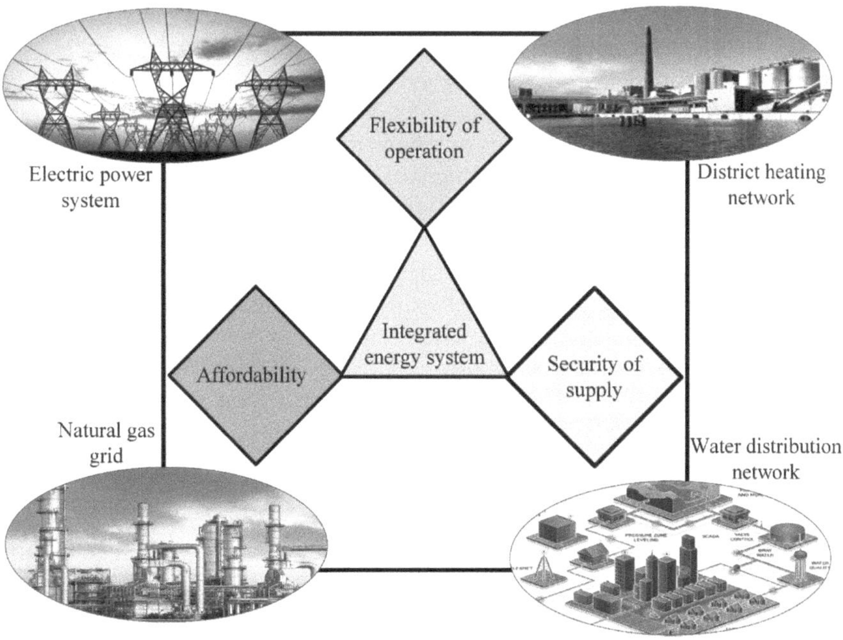

FIGURE 2.5 Three key advantages of the multi-carrier energy network.

Hence, there are undeniable dependences among multi-carrier energy grids that need to be intended in the optimal operation of the grid [15]. On the other hand, great advances in the technology of appliances have been led to the development of new hybrid energy systems that have the capability of working with diverse carriers of energy. Such systems can simultaneously input uncertainties in the consumption of various carriers of energy affecting the optimal operation of the related network [16]. Thus, modern energy networks face interdependences between several energy grids and multi-energy interactions that occur among energy carriers such as electrical, gas, heating, cooling, and water energy [17]. This issue inspires the inevitable fact of integrating different energy grids and creating the multi-carrier energy network with the aim of effectively modeling interdependences and multi-energy interactions among several hybrid energy units [18]. Fig. 2.5 indicates three key advantages of the multi-carrier energy network.

According to Fig. 2.5, one of the attractive benefits of multi-carrier energy networks is to increase the flexibility of the operation for the integrated grid [19]. Indeed, when different energy grids are operated in the interconnected format, it allows the whole system to effectively use the potential of various networks for supporting other ones when they are in abnormal conditions. For example, as FMENs are targeted to equip full RESs for carbon-free energy production, they may face produced energy larger or lower than the forecasted one. In such circumstances, the interconnectedly exploitation of the system enables it to use the capability of energy conversion and storage technologies to receive (or transmit) required energy from other grids and keep the continuous energy supply in the system. In other words, coupling various grids adds another promising way for flexibility of the system and allows it to behave more flexibly than an independent state. This property also makes the system susceptible to optimal planning with minimum carbon emissions and operation costs and can support it for optimally utilizing the available assets. However, it can also bring critical challenges when unexpected malfunctioning of coupling components leads to the unpredicted potentially negative impacts of grids on each other. This issue can result in losing energy as well as increasing carbon emissions and operation costs.

Affordability is another key advantage of the multi-carrier energy network as it can provide cost-effective operation of the system [20]. Achieving this benefit is mostly back to the capability of the system to effectively avail of grids' capacities for the exploitation of the system with minimum costs. As an example, when inadequate gas sometimes leads to increasing its price and shortage, the integrated structure can use the excess of produced electricity and conversion facilities to convert power to gas for supporting the gas system to affordably meet multi-energy. Indeed, effective usage of RESs potential, capturing and converting produced clean energy via multifarious coupling components can strengthen the ability of the system for affordable operation. In this regard, the security of supply is another benefit of coupling diverse energy grids into the united structure. The integrated structure allows the system to have several options for supplying multi-energy. Indeed, the coupled grid is able to utilize several available coupling components for supporting uninterrupted multi-energy serving using the capacity of the whole interconnected grid. This ability makes supplying a certain carrier of energy possible when the system suffers from its shortages by converting other carriers of energy to it. Hence, the integrated energy grid acts as a promising way for improving the security of multi-energy supply.

2.5 Summary

The ever-increasing energy crisis alongside rapid developments in technology has created an urgent need for a revolution in the energy landscape. Due to this, this chapter is aimed to provide a complete assessment regarding the requirements for grid modernization. To this end, the grid revolution was investigated from three main aspects including residential, industrial, and multi-carrier energy network aspects. The key elements affected in modernizing the residential side were discussed to give a clear overview for transiting towards the modern residential sector. The four transformations in the industrial sector were analyzed from industry 0.1 to 0.4 aiming to highlight the industry part is targeted to include which attributes. The network expansion was explored from various viewpoints to clarify why modernization needs to be considered in the energy grid expansion. Additionally, why future energy networks need to be a coupled version of different grids was assessed and the advantages of integrated energy systems were also described to justify the interconnecting of energy grids.

2.6 Questions

The careful analyzing whole of this chapter is expected to enable readers to effectively evaluate the following questions.

1) Why the modernization of energy networks is necessary at the whole level?
2) What are the key requirements for switching from traditional grids to modern energy networks?
3) Which parts of the energy network can have the most impact from grid modernization?
4) Why does the residential-side need to be modernized and how it can be possible?
5) What are the prominent requirements for modernizing the residential sector?
6) What are the impacts of grid modernization on the residential sector?
7) What are the developments in the industry sector from industry 0.1 to 0.4?
8) What are the goals for the industrial revolution in the future years?
9) Why the grid modernization is an essential step for future industry evolutions?
10) What are the challenges for achieving the modernization goals of the industry sector?
11) Which factors need to be intended in the energy network expansion?
12) Which parts of the energy network can have the most impact from grid modernization?
13) What are the processes in the three main parts of the energy network?
14) Why do the different energy networks need to be interconnectedly operated?
15) What are the advantages of integrated energy grids?
16) The hybrid system can face which challenges in the grid modernization process?
17) Why does the multi-carrier energy network need to be intended in modernizing the future energy grid?

2.7 Suggestions

Given the topic of this chapter, the following suggestions are provided for the readers assisting them to pursue related research clues.

1) Analyzing the key requirements for energy network modernization can facilitate the successful realization of grid modernization goals.
2) Identifying the effective requirements for modernizing the residential side is one of the essential steps of grid modernization.
3) Developing a holistic architecture can support the modernization of the industrial sector.
4) Considering grid modernization criteria in the energy network expansion is necessary for modernizing future energy grids.
5) Developing hybrid paradigms is a key requirement for modernizing future multi-carrier energy networks.

2.8 Future trends

Future trends can give a proper overview for the readers to form their related works and develop effective solutions for remaining gaps, which are listed as follows for this chapter.

1) All requirements for successful switching from traditional grids to modern energy networks need to be identified and evaluated.
2) The need for modernizing the residential side should be investigated from different perspectives.
3) Comprehensive models need to be developed for modernizing the industrial sector.
4) The modernization criteria need to be expanded for modernizing energy grids.
5) Hybrid paradigms need to be designed for facilitating the modernization of future multi-carrier energy networks.

References

[1] A.S. Meliopoulos, G.J. Cokkinides, R. Huang, E. Polymeneas, P. Myrda, Grid modernization: seamless integration of protection, optimization and control, in: 2014 47th Hawaii International Conference on System Sciences, IEEE, 2014, pp. 2463–2474.

[2] M. Daneshvar, B. Mohammadi-Ivatloo, K. Zare, M. Abapour, S. Asadi, A. Anvari-Moghaddam, Chance-constrained scheduling of hybrid microgrids under transactive energy control, Int. J. Energy Res. 45 (7) (2021) 10173–10190.

[3] A. Pratt, D. Krishnamurthy, M. Ruth, H. Wu, M. Lunacek, P. Vaynshenk, Transactive home energy management systems: the impact of their proliferation on the electric grid, IEEE Electrification Magazine 4 (4) (2016) 8–14.

[4] Y. Wu, Y. Wu, J.M. Guerrero, J.C. Vasquez, A comprehensive overview of framework for developing sustainable energy internet: from things-based energy network to services-based management system, Renew. Sust. Energy Rev. 150 (2021) 111409.

[5] Y. Cao, et al., Development and transient performance analysis of a decentralized grid-connected smart energy system based on hybrid solar-geothermal resources; Techno-economic evaluation, Sustain. Cities Soc. 76 (2022) 103425.
[6] G. Pinto, Z. Wang, A. Roy, T. Hong, A. Capozzoli, Transfer learning for smart buildings: a critical review of algorithms, applications, and future perspectives, Advances in Applied Energy (2022) 100084.
[7] G. Chicco, P. Mancarella, A unified model for energy and environmental performance assessment of natural gas-fueled poly-generation systems, Energy Convers. Manage. 49 (8) (2008) 2069–2077.
[8] L. Cui, S. Weng, A.M. Nadeem, M.Z. Rafique, U. Shahzad, Exploring the role of renewable energy, urbanization and structural change for environmental sustainability: comparative analysis for practical implications, Renew. Energy 184 (2022) 215–224.
[9] S. Katipamula, J. Haack, G. Hernandez, B. Akyol, J. Hagerman, VOLTTRON: an open-source software platform of the future, IEEE Electrification Magazine 4 (4) (2016) 15–22.
[10] (March 9, 2022). *U.S. residential sector energy consumption by energy source*. Available: https://www.eia.gov/energyexplained/use-of-energy/homes.php#:~:text=Electricity%20is%20used%20in%20almost,use%20energy%20consumption%20in%202020.
[11] M. Daneshvar, B. Mohammadi-Ivatloo, K. Zare, A. Anvari-Moghaddam, Transactive Energy Strategy for Energy Trading of Proactive Distribution Company with Renewable Systems: a Robust/Stochastic Hybrid Technique, e-Prime-Advances in Electrical Engineering, Electronics and Energy (2022) 100028.
[12] M. Daneshvar, B. Mohammadi-Ivatloo, K. Zare, A novel transactive energy trading model for modernizing energy hubs in the coupled heat and electricity network, J. Clean Prod. (2022) 131024.
[13] X. Dong, J. Wu, Z. Xu, K. Liu, X. Guan, Optimal coordination of hydrogen-based integrated energy systems with combination of hydrogen and water storage, Appl. Energy 308 (2022) 118274.
[14] M. Daneshvar, B. Mohammadi-Ivatloo, S. Asadi, M. Abapour, A. Anvari-Moghaddam, A transactive energy management framework for regional network of microgrids, in: 2019 International Conference on Smart Energy Systems and Technologies (SEST), IEEE, 2019, pp. 1–6.
[15] T. Gronier, J. Fitó, E. Franquet, S. Gibout, J. Ramousse, Iterative sizing of solar-assisted mixed district heating network and local electrical grid integrating demand-side management, Energy 238 (2022) 121517.
[16] M. Daneshvar, H. Eskandari, A.B. Sirous, R. Esmaeilzadeh, A novel techno-economic risk-averse strategy for optimal scheduling of renewable-based industrial microgrid, Sustainable Cities and Society 70 (2021) 102879.
[17] A. Shabanpour-Haghighi, M. Karimaghaei, An overview on multi-carrier energy networks: from a concept to future trends and challenges, Int. J. Hydrogen Energy (2021).
[18] G.T. Udeh, S. Michailos, D. Ingham, K.J. Hughes, L. Ma, M. Pourkashanian, A modified rule-based energy management scheme for optimal operation of a hybrid PV-wind-Stirling engine integrated multi-carrier energy system, Appl. Energy 312 (2022) 118763.
[19] M. Daneshvar, B. Mohammadi-Ivatloo, K. Zare, S. Asadi, Transactive energy management for optimal scheduling of interconnected microgrids with hydrogen energy storage, Int. J. Hydrogen Energy 46 (30) (2021) 16267–16278.
[20] G. Sarlak, J. Olamaei, M. Dosaranian-Moghadam, Optimal Management of Multi-Carrier Energy Hub Regarding Fuel Cell and, Storage Technologies, J. Energy Storage 46 (2022) 103821.

CHAPTER

3

Energy trading solution: The capable leverage for a renewable-dominant future

3.1 Introduction

Recent years are accompanied by exceptionally advances in energy networks' paradigms that affect the system's operation from the local level to the network range [1]. In addition to this, the current structure of energy networks is undergoing sophisticated evolutions by adopting innovative technologies [2]. Such transformations are particularly for accommodating the ever-increasing penetration of renewable systems and new emerging loads, especially those ones for the decarbonization objectives [3]. The technology of the wind and solar systems has witnessed accelerated development due to economic incentives, supportive energy policies, and favorable environmental outcomes [4]. However, the dependency of renewables' outputs on climate changes has added uncertainties to the optimal scheduling of the system, which inspires the need for confident ways of energy supply. One of the appropriate ways of dealing with this challenge is to deploy the different types of energy storage technologies that offer substantial operational advantages [5,6]. Nevertheless, another side of the usage of energy storage technologies is their high costs imposed on the system. The integration of distributed energy resources (DERs), especially renewable types of them is another serious solution for the high usage of intermittent energy resources [7,8]. However, this solution is also faced the key question of how the integration of renewable systems can be optimally realized in a sustainable manner. In this respect, recent advances in technology, communication, and computation have facilitated the widespread and massive integration of decentralized energy systems into the energy grid.

Emerging Transactive Energy Technology for Future Modern Energy Networks
DOI: https://doi.org/10.1016/B978-0-323-91133-7.00010-7

This progress does not alone come with tremendous opportunities and benefits whereas basic challenges arise for stakeholders such as policymakers, retailers, residential, commercial, and industry users [9]. It requires a comprehensive framework to make the optimal coordination of numerous DERs agents possible in the presence of several intelligent devices. In continuing of developing innovative ways for the integration of renewable energy resources (RERs), energy trading mechanisms have appeared as promising solutions for easing the energy interactions throughout the energy structure [10,11]. The idea of establishing transactions for energy backs to the necessity of finding affordable ways for managing energy in the deregulated environment. Energy trading mechanisms can enable different players to act as energy sellers or buyers during various time periods of the day. Energy trading paradigms are more adaptable with systems that are highly equipped with renewable energy units. Indeed, the uncertain behavior of RERs in energy generation makes the space susceptible to having different types of energy transactions during the day. This issue indicates the importance of energy trading strategies in doing a central role in the energy management of the grid. This chapter proposes an energy trading solution as the capable leverage for the optimal integration of RERs in the future multi-energy network. For this aim, it investigates the developed energy trading-based models in recent studies from different viewpoints. The detailed analysis indicates the vital role of energy trading in enabling the related model in achieving the targeted goals. It also includes the assessment of various types of energy-sharing models that highlight their strongness and weakness. Such assessments procure a helpful overview regarding energy trading strategies and guide the decision-maker to adopt appropriate ones in line with the requirements of the related system.

3.2 Energy trading models in recent studies

3.2.1 Operation studies

Given the mentioned important role of energy trading in highly utilizing RERs for energy generation, the energy trading-enabled models are attracted more attention in recent works. From the operational viewpoint, different optimization models are developed for the effective usage of energy trading possibilities in the studied system. Reference [12] proposes a collaborative operation framework with Peer-to-Peer (P2P) energy sharing for promoting the interactive efficiency among hydrogen fueling stations and the power grid. The model is proposed considering the notable interactions between the transportation network and the power network due to the prevalence of hydrogen fueling electric vehicles. Given the provision of a promising solution by the P2P transactive energy strategy for the secure and efficient operation of the distribution network, the authors in [13] offered a security-constrained decentralized P2P transactive energy sharing model for allowing direct energy trading between neighboring prosumers. The Nash Bargaining theory is used to formulate the P2P transactive energy trading problem by employing the decomposition method. The emerging communication and information technology along with enhancing the popularity of decentralized energy units have brought new opportunities with various platforms for energy exchange among prosumers that were already consumers. Thus, the authors proposed a distributed carbon-aware energy trading mechanism in [14] to coordinate prosumers in the virtual power plant. A novel stochastic optimization framework is presented in [15] to resilient operate the interconnected energy hubs using the P2P energy trading for creating energy sharing possibilities among them during severe disturbances. The optimization model is completed by considering the duration of disturbances as well as common uncertainties of renewable energy units. Considering the current trend of distribution systems that are moving towards the distributed form of operation with a high level of RERs, reference [16] proposed a new decentralized energy market framework enabled by P2P energy trading. The market structure is formed by participating P2P agents along with the nodal agents and the alternating direction method of the multiplier algorithm is applied for achieving its equilibrium. A transactive energy scheme is presented in [17] for providing direct P2P energy sharing among prosumers. In this work, coordinating with neighbors was the main criteria for prosumers to share ideal energy for maximizing their benefits. The uncertainties in renewable-based systems have brought the necessity of probabilistic modeling of the related optimization problems. Due to this, a data-driven distributionally robust co-optimization model is proposed in [18] for the network operation of interconnected microgrids considering the P2P energy sharing among them. The emerging soft open point is applied to flexibly connect multi-microgrids for the implementation practices. Herein, maintaining the operational security of the distribution network is still remained challenging for distribution grids with increasing penetration of RERs. Given the possibility of P2P energy transacting among small-scale consumers of such systems, a novel integrated operation and transaction model is proposed in [19] to establish P2P energy sharing as well as ensure the security constraints of the distribution grid. To solve the conflict of interests of prosumers in the energy tradings, the Vickrey-Clarke-Groves mechanism is applied considering the operational requirements for the distribution grid. Reference [20] proposed a supervised-learning based aggregated exploitation model of different heterogeneous DERs with the aim of estimating the aggregated parameters in P2P energy trading. In this work, the model is empowered by utilizing the fast alternating direction method of multipliers for accelerating the iteration process among aggregators. A Markov decision process is considered in [21] to formulate an energy cost minimization problem for the participated prosumers in the P2P energy sharing market. A multi-agent deep deterministic policy gradient algorithm is also proposed for supporting the system in learning optimal energy trading decisions and benefiting an efficient and accurate energy trading model. A game-theoretic model is proposed in [22] for P2P energy trading of the prosumer and a consumer in the energy community. The Stackelberg game is used to form a real-time pricing model, in which prosumer and consumer respectively play as leader and follower in the related game. The P2P energy exchanging model is designed in [23] by deploying transactive energy technology to enable prosumers for energy trading in the community that most of its capacity is fulfilled with renewable systems. In this work, economic, technical, and environmental objectives are intended for cooperative prosumers in the deregulated environment. In another study [24], the existence of distributed P2P energy trading market was on the agenda due to the continuous expansion of DERs on the energy demand side. A bilateral energy sharing structure is proposed based on the energy trading agent with an energy storage system and self-built photovoltaic (PV) with the aim of procuring more sufficient energy exchanges among the community of prosumers. A cooperative operation framework model for decentralized multi-energy generation units is presented in [25] to cope with the challenge of not mostly considering renewable systems in the conventional operational strategies. Indeed, the Nash Bargaining optimization problem is intended for the optimal operation of multi-energy systems to maximize social welfare instead of the traditional non-cooperative solution-based approach. In [26], the energy trading strategy is expanded to the proactive distributed company to allow it for the optimal operation in the presence of different types of DERs. In this extension, the transactive energy paradigm is applied to give the energy trading possibilities for the components in the distribution system. Moreover, in the employed operation strategy, the hybrid robust stochastic method is considered for dealing with uncertainties of various uncertain parameters in the energy structure. In another research [27], the management of a more proactive energy demand-side is intended along with the wide deployment of RERs to focus on the new paradigm in both electricity market trading as well as the power system operation. Due to this, the authors presented a novel P2P power trading framework using the generalized fast dual ascent approach considering the network security constraints aiming to reach the flexible local market mechanism for properly responding to fast changes of RERs. Dealing with the rapid changes of RERs is considered in [28] by proposing a robust-based model for the optimal operation of distribution networks. In this article, the P2P transactive energy trading framework is investigated to allow prosumers for energy sharing to solve significant challenges ahead of the system's operation. Furthermore, a vertex scenario-based robust optimization technique is proposed to handle uncertainties and offer the robust condition for the exploitation of the distribution grid. The transactive energy application is extended in [29] for the energy management of interconnected microgrid clusters. In this study, the model is empowered by the chance-constrained method for taking the stochasticities of uncertain parameters into account in the system modeling. In [30], transactive energy trading is considered for rural network electrification by adding energy sharing possibilities to the energy management options. The residential and commercial microgrids are scheduled by optimally operating DERs to minimize the overall cost of the system. The energy trading solution is intended in [31] as the control way of multiple microgrids based on transactive energy management. Indeed, this work focused on using the transactive energy exchanging to optimally exploit decentralized energy systems for community microgrids. The new study in [32] relies on changing the conventional utility-dominated tariff-based business model and transforming it into the emerging P2P energy trading model due to no longer suitability of it for the highly penetrated DERs systems. For this aim, this research proposed a two-stage P2P energy sharing model by taking the utility role into account in solving the transaction approval model for verifying the energy transactions from the secure system exploitations viewpoints. Another work [33] intended the ownership complexity of DERs when their integration into community microgrids is widely promoted due to extending the carbon reduction schemes across the globe. A game-theory-based decentralized energy sharing program is developed to analyze DERs ownership effects on the participants' benefits in the P2P energy exchanging market with direct energy trading among microgrids without any intermediary. The integration of distributed RERs is assessed in [34] by considering their intermittences as the challenge for the efficient exploitation of residential distribution networks. The paper proposed a P2P energy exchanging platform as a promising solution for tackling the mentioned challenge to coordinate demand-side energy management programs and establish energy transactions for residential houses. P2P energy sharing is employed for smart homes in [35] considering the interactive visual interface and responsive demand for monitoring of the system. This study created an active market structure for the operation of the system when it benefits P2P energy trading for regulating optimal energy interactions. In another work [36], the application of energy trading is used for congestion management in the deregulated energy market structure. For this aim, the authors suggested a new approach for considering the time sequence of P2P energy sharing under the continuous double auction mechanism to solve the physical congestion of the energy grid. Given the participation of electric vehicles (EVs) in the energy trading interactions, the authors in [37] proposed an operational P2P framework for the energy exchange of EVs within a charging station and a business entity with PV for clean energy production. This study develops a dynamic pricing mechanism for EVs with the aim of increasing owners' profit as well as promoting the contribution of stations for charging EVs in the P2P energy sharing market. In order to ensure the network constraints are maintained in the allowable range as a major challenge of the system with energy sharing possibilities, a new methodology is proposed in [38] based on the sensitivity analysis to examine the impact of P2P exchanges on the grid and guarantee keeping network constraints during energy sharing. A competitive market consisting of retailers and prosumers is designed in [39] to allow them to conduct P2P energy sharing in the system. A primal-dual sub-gradient algorithm is used in this work as a novel decentralized method for clearing the structured market without involving third-party or disclosing the private information of market players. The transactive energy application is extended in [40] for the optimal dispatch of integrated low-carbon energy hubs in the gas and electricity incorporated networks. In this study, the model is empowered by the stochastic programming methods for taking the stochasticities of uncertain parameters into account in the system modeling. The transactive energy trading framework is suggested in [41] for the power and gas energy management in the coupled electricity and gas energy grids. This work also completed the optimization model by using the distributionally robust chance-constrained approach for the uncertainty quantification in the probabilistic modeling of the system. In another work [10], the optimal operation of hybrid microgrids is conducted considering multi-energy interactions in the interconnected energy network. Moreover, hydrogen energy is involved in the energy management of the system in [42] when the transactive energy trading technique is applied to establish energy exchanging opportunities for the interconnected renewable-based microgrids. Some of the prominent objectives of applying energy trading solutions to the energy network's operation problems can be summarized as follows.

- Improving energy interactions in the system;
- Secure and efficiently operating the distribution network;
- Supporting resiliency schemes against severe disturbances;
- Ideal energy sharing for pursuing different objectives;
- Easily satisfying the network constraints;
- Facilitating the adoption of a high or full level of RERs;
- Coordinating demand-side energy management programs;
- Managing the physical congestions of the grid;
- Energy market clearing without additional parties.

3.2.2 Planning studies

As renewable systems are day by day becoming famous energy generation units for their capabilities for clean energy production, the flexibility in energy supply is being a prominent factor with a similar rate in energy system modeling. Energy trading is one of the promising tools for satisfying the flexibility requirement of grids with a high penetration of RERs. Energy trading applications as the solution for energy management are not just restricted to operation problems but also their usage is moved far more extensive than operation level and widely intended for the energy system planning studies. Reference [43] proposed a bi-level robustly game-based planning framework to model bilateral transactions among microgrids and the distribution network in the emerging power market. The effects of bilateral power trading are intended between multiple stakeholders in the energy planning problem. The P2P energy trading is taken as the main tool in [44] for the planning of the grid-connected microgrids under the advanced optimization model. A game theory method is applied to determine the Nash equilibrium and structure the multi-objective optimization model with the aim of finding optimum payoff values and correct sizing of DERs. A multi-period P2P energy trading model is proposed in [45] for indicating the economic viability of inter-entity power planning and creating horizontal cooperation between entities for allowing internal power sharing. This work also adopted a fuzzy optimization method with the aim of ensuring simultaneous optimization of the conflicting objectives without over-prioritizing of them under the fuzzy sets theory. In [46], the authors presented a novel bi-level transmission expansion planning model to facilitate energy trading among energy producers and consumers as well as minimize the grid investment cost. A collection of market clearing problems at the lower-level constraints the upper-level problem that represents pool trading and individual objective functions related to social welfare. A distribution network planning model is developed for the utility in [47] to enable end-users to directly energy exchange considering neighborhood energy sharing platform. The optimal decisions are provided for the transformer and line upgrades as well as battery energy storage system and PV panels in end-users premises under the suggested mixed-integer second-order cone programming model. In [48], an optimal reinforcement planning strategy is proposed to procure economic and reliable energy services for the consumers given the complicated status of the distributed systems management due to introducing the new energy-sharing technologies and strategies. Moreover, distributed generation systems are taken into account as a viable option for the system's reinforcement in the newly deregulated energy market environment. The impact of energy sharing between neighbors on the planning and operation of the distribution network is analyzed in [49] with the aim of creating an energy balance between energy supply and demand. This investment is carried out for the energy community equipped with DERs like PV and storage systems to explore their benefits in improving the energy efficiency of the grid. Path planning is suggested for energy management in [50] that is structured based on the blockchain technology for efficient energy sharing between charging stations and EVs. A game theory-based solution is also offered to support peak load control and energy management in the peak and off-peak conditions of the system. The energy planning model is developed in [51] for net-zero energy communities using the P2P energy sharing management and advanced storage units considering data-driven machine learning predictions, stochastic uncertainty sampling, and transient system modelling based on climate changes by 2050. The large renewable energy production is intended in developing the P2P energy sharing management strategy to determine time-of-use energy periods with the aim of improving energy flexibility. The planning of the networked microgrid is performed in [52] by proposing a multi-objective technique based on the peer-to-grid and P2P energy exchanging schemes. Energy index of reliability and annual profit are considered as two various criteria for forming the multi-objective function and a cooperative game theory method is used based on the particle swarm optimization algorithm for finding the appropriate size of the players as well as maximizing the payoff values of the objective functions in modeling the networked microgrid. Reference [53] presents new interoperable functionalities for online planning and stochastic two-way energy exchanging to improve both productivity and efficiency. This research benefited from the Lyapunov optimization method for addressing the challenges and uncertainty optimization of energy planning and sharing. Reference [54] proposed a new business model for optimal planning and operation of community microgrids by developing a frame for a flexible retail power market. The microgrid is equipped with different types of decentralized energy generation units such as micro diesel engines, battery storage systems, and renewable energy units to minimize the operational costs of the overall system. The day-ahead planning of energy systems is done in [55] at trading sections of energy companies. This assessment is conducted considering the energy trading company as the decision maker and assuming that the energy supply and demand are certain in advance. An energy production planning problem is formulated in [56] for a virtual power plant considering different energy generation constraints. The virtual power plant is targeted to provide an energy trading possibility for participating in the energy market by aggregating distributed power producers, dispatchable loads, and energy storage units, which facilitate pursuing different objectives of the energy market's participants. Reference [57] presented an investment planning scheme to improve the profit of the distribution network company. The role of the P2P energy trading strategy is investigated as an attractive and profitable option for energy companies to effectively use DERs for following technical and economic objectives. This research also introduced network usage charges for procuring charging services and optimizing the distribution network planning by focusing on the power transactions. A combined power and gas energy market mechanism is presented in [58] for allowing two-stage energy planning and sharing based on asynchronous gas and power markets. This work scrutinized the interdependencies between power and gas energy systems by clearing the gas market to maximize social welfare in the first stage problem while maximizing corresponding profits of the energy generation and consumer companies in the planning horizon of the second stage optimization problem. To reduce the risk of planning under the deterministic techniques, a stochastic optimization technique is proposed in [59] for the generation and trading planning of the system with the aim of maximizing the total utilities' income. The model for the complicated problem decomposes the whole system into subsystems for driving the system to get the optimal solutions using the energy exchanging opportunities considering various types of uncertainties in the deregulated energy market. Some of the prominent objectives of applying energy trading solutions to the energy network's planning problems can be summarized as follows.

- Creating horizontal cooperation between energy entities
- Enabling end-users for directly energy exchanging
- Procuring economic and reliable energy services
- Supporting peak load control and energy management
- Realizing net-zero energy communities
- Improving energy flexibility for energy networks
- Improving both productivity and efficiency for energy systems
- Facilitating the effective usage of various types of DERs
- Reducing the risk of energy network planning

3.3 Renewable-dominant future energy networks

Future energy grids are planning to serve clean, efficient, and reliable energy to energy consumers [60]. Completely reaching such goals faces critical challenges in switching from traditional networks to modern energy infrastructure [61]. Already and current structure of energy grids mostly depends on the centralized power generation units that bring different threats such as environmental and economic criticisms [62]. However, with developing technologies and widespread advances in energy systems, such models of energy production are no longer acceptable in the energy grid expansion planning. Hence, the trend is going towards decentralized clean energy production systems with RERs as popular energy sources, especially for the future [63]. Similar trends are formed to utilize the

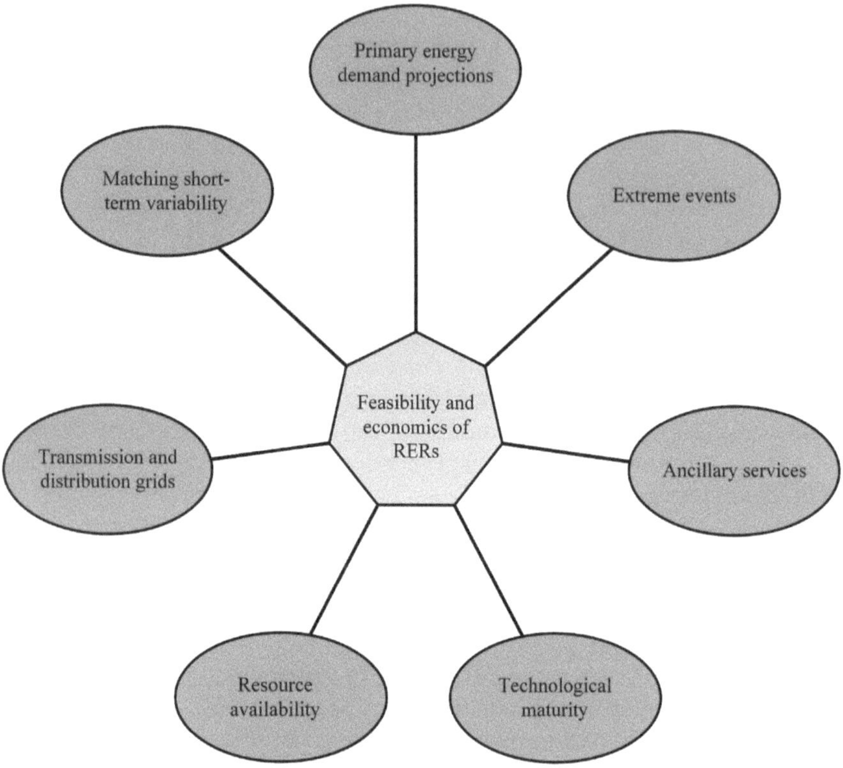

FIGURE 3.1 Some prominent factors for the feasibility and economics of RERs [66].

advantages of RERs in the energy infrastructure and then made RERs as the dominant sources of energy for future energy networks [64]. Clean energy output without requiring other costly external fuels for energy production has converted RERs as an affordable option for extending energy networks, particularly in the energy production section [65]. Indeed, RERs offer environmentally friendly energy generation processes and are economically viable with respect to the different factors. Some prominent factors are illustrated in Fig. 3.1 [66].

As RERs provide substantial benefits for energy networks, using these uncertain energy generation units inputs a massive volume of stochasticity into the optimal energy systems operation [67]. Indeed, this is one of the main challenges in using RERs in the network that calls the research world for practical solutions applicable to such problems. As future energy grids are planned to equip with 100 percent RERs [68], this challenge finds a high value for the energy network expansion. For reaching such a key goal, different research works are carried out by concentrating on developing practical solutions for the feasibility of 100 percent RERs plans. In [69], the interconnection of power networks with 100 percent RERs is investigated to analyze the achievement of greater economic benefits by attending to the existence of the technology for long-distance, high power Ultra High Voltage Direct Current transmission. This work also presented sensitivity analyses with respect to diverse weather years and infrastructure capital costs that indicate an approximately 18 percent reduction for the system in the total annual cost when it is interconnectedly operated. A real complex power grid with 100 percent RERs is scrutinized in [70] aiming to focus on the primary frequency regulation. In this investigation, an electromechanical model is used to represent the power generation park and the grid in a free open-source area. Moreover, the control system of doubly fed induction generators is evaluated considering the implementation of a new control strategy with the aim of avoiding maximum power point tracker exploitation. In [71], a novel operational model is proposed for the reliable and secure operation of community microgrids that are equipped with 100 percent RERs for carbon-free power production. The proposed model stands on the new energy trading strategy for energy exchanging of microgrids in the transactive energy market resulting in dynamically balancing energy between supply and demand while realizing the economic and technical objectives. A hybrid information gap decision theory and stochastic programming method is also availed for driving the decision maker for purposeful decisions in the uncertainty quantification process. The energy availability and cost of energy systems are analyzed in [72] by considering the aim of adding biomass power to 100 percent RERs of Akita prefecture. Hourly energy demand and RERs output are explored to estimate the differences in the energy generation and consumption, which the battery is intended for supplying the energy shortage brought from the inadequate power production by RERs. A novel modeling framework is proposed in [73] for multi-year energy production with the perspective toward 100 percent RERs and transmission planning problems with the aim of minimizing both operational and expansion costs under the continuous and binary uncertainties. The combined robust optimization technique and stochastic programming method is developed in the form of a decomposition-based solution strategy to capture multiplex uncertainties in the coordinated long- and mid-term planning of the power system in the presence of 100 percent RERs. Reference [74] developed a new model that enables microgrids to benefit from 100 percent clean energy production by equipping them to fully RERs in response to the significant challenges caused by conventional systems. In this work, a two-stage stochastic robust programming model is proposed to not only handle the impacts of multifarious uncertainties but also to maximize the expected profits of microgrids through the optimization of their bidding strategy and minimize the energy imbalance cost by adjusting DERs in the real-time market. The authors in [75] focused on evaluating a detailed expansion planning for the energy production, transmission, and storage to reach the goal of 100 percent RERs by 2030. This work used Balmorel as an optimization tool for optimizing dispatch and investments and also developed a new method for translating optimal results to a realistic RoadMap. Reference [76] aimed to examine the use of current engineering and scientific theory and practice for refuting the principal myths. This study performs the mentioned investigation by considering 100 percent renewables for large-scale electricity systems that make them variable to design appropriate solutions for meeting the key requirements of affordability, security, and reliability. In [77], the operation of electric power systems is analyzed by intending the structure with extremely high levels of variable RERs for going towards achieving 100 percent renewable energy networks. Simulations for 100 percent renewable energy units are presented in [78] to assess meeting the actual hourly power load in one territory and five states spanned by the Australian National Electricity Market. This study is done by considering commercially available technologies like existing bio-fueled gas turbines and hydropower, PV panels, wind turbines, and solar thermal power with the thermal storage unit. Reference [79] used two-stage mixed methods to explore the implementation and adoption of 100 percent RERs for different cities in the U.S.A with the aim of assessing their needs for new administrative capacities and institutions to successfully implement the related plans and policies. For this aim, multivariate regression is conducted to examine policy factors, institutional, and demographic in 100 percent RERs adoption. The authors outlined why and how 100 percent fluctuate renewable energy systems can be integrated by taking a Smart Energy Systems method in [80]. Indeed, the design and development of coherent Smart Energy Systems are presented as an interconnected part of reaching the future 100 percent RERs and transport solutions to investigate the creation of the flexibility necessary for the integration of large penetrations of intermittent RERs. The economic viability and technical feasibility of 100 percent renewable energy units are explored in [81] that includes desalination, heat, power, and transport sectors for a global megacity Delhi. This research presents a cost-optimal, multi-regional, multi-sectoral, and technology-rich energy transition pathway for driving it away from fossil fuels towards RERs for Delhi as a hub within the regional energy units. In another study [82], the possibility of reaching to 100 percent RERs is explored for Sweden by using practical data and replacing current fossil fuel-based thermal generation systems with wind power producers. The results of this research indicated that the Swedish power production system can achieve 100 percent RERs by tripling the current wind power production capacity and integrating with the existing hydropower systems in the country. A similar study is done for Austria in [83] with the goal of analyzing the possibility of meeting total energy consumption using RERs while dynamically balancing energy, preserving the security of energy

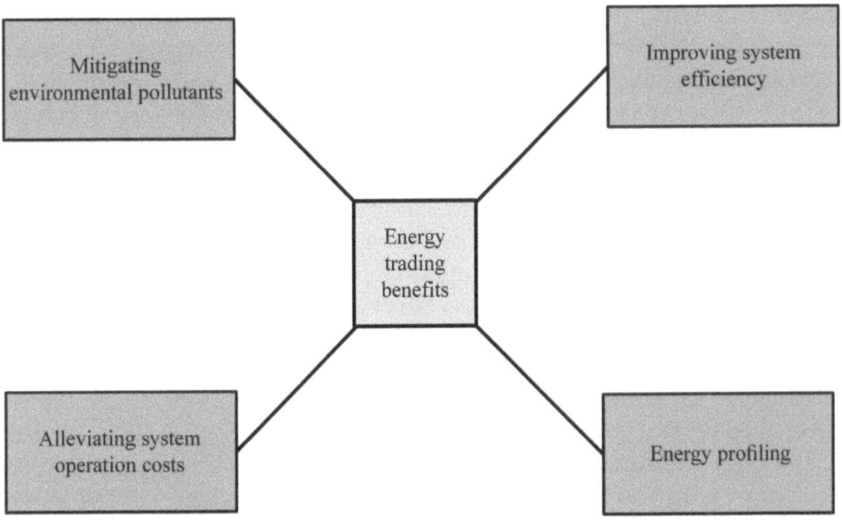

FIGURE 3.2 Typical benefits of energy trading solutions [84].

supply, and safeguarding system stability. According to this assessment, hydro, PV, and wind are responsible for a high share of clean electricity generation that input remarkable weather-related fluctuations to the energy supply process requiring flexibility improving solutions for reliable energy serving.

3.4 Energy trading mechanisms

In the context of high usage of RERs in future energy grids, different solutions are suggested to deal with the uncontrollable features of RERs that energy storage systems, optimal integration of RERs, and demand-side energy management programs are some of the important ones. In addition to the mentioned ways of easing RERs adoption, energy trading methods are also recognized as promising solutions for doing such responsibilities. Energy trading solutions refer to the sharing mechanisms that allow energy market members to buy or sell a certain amount of energy based on different strategies. The motivations for energy exchange can vary from one participant to another participant and it can be typically technical, environmental, and economical goals. Meanwhile, energy trading solutions bring some typical benefits that are demonstrated in Fig. 3.2 [84].

According to Fig. 3.2, one of the important benefits of energy trading mechanisms is the mitigation of environmental pollutants. This can be occurred by easing the adoption of high-level RERs for energy production instead of relying on fossil fuel-based units. Energy trading strategies can improve the efficiency of the system by creating a new way of energy interactions for the grid members that support the system in effectively balancing energy [85]. Using energy-sharing opportunities for the energy management of the grid can reduce the need for extra capacities of energy storage systems and power production units resulting in alleviating the operation costs of the studied system. Another key benefit of energy trading solutions is energy profiling. Indeed, energy exchanging mechanisms can open a new door for the grid to manage energy interactions to effectively establish energy balance between energy supply and demand in all time periods of the day. In addition to the mentioned usual benefits of the energy sharing mechanisms, all these mechanisms have their special own advantages for the participants that allow them to suitably pursue their diverse objectives in the energy network.

Energy exchanging is a key and basic action for any energy-sharing mechanisms in energy interactions. However, how these energy transactions need to be conducted is an important issue that backs to the type of energy sharing mechanisms. Due to this, various energy trading mechanisms and techniques are developed considering the type of market and prominent factors for the participants. Some of the important of these mechanisms are described in the following subsections.

3.4.1 Real-time energy trading mechanism

In systems with a high/full share of RERs, establishing a time-to-time energy balance is critical for keeping the stability of the energy grid. Herein, a real-time energy trading mechanism is one of the promising ways for making the energy environment susceptible for eligible participants to trade energy for creating a real-time energy balance in the energy infrastructure. Reducing payment for the energy bill, procuring a higher degree of flexibility, decreasing the operation costs, and enhancing the quality of the energy supply are some of the advantages of the real-time energy trading mechanism [86].

3.4.2 Peer-to-peer energy trading mechanism

In energy networks, prosumers are entities that are able for both energy producing and consuming while consumers can only consume energy and then will be energy buyers in the energy market. Both prosumers and consumers can have energy trading with each other for meeting their energy requirements and following their goals. In the energy world, the direct energy sharing between them is called P2P energy exchanging [87], which is developed under the concept of sharing economy [88]. A peer of the P2P energy trading mechanism refers to a group or one market member that can directly sell or purchase multi-energy with a corresponding peer without intermediating of the external entity [89]. The P2P energy trading mechanism is widely applied to different energy problems and provides a wide area of benefits for the sustainability of future energy networks with a high/full penetration of RERs.

3.4.3 Blockchain-based energy trading mechanism

One of the secure structures for energy trading can be built based on the blockchain technology that allows for the usage of smart contracts in the P2P energy exchanging market by actualizing the automated execution. Blockchain technology can offer effective solutions for different issues from the coordination of RERs portfolios and energy storage management to demand response services [90].

3.4.4 Edge computing-based energy trading mechanism

Despite privacy and security being two key factors in the problems associated with blockchain technology, the edge computing mechanism has appeared as an essential tool for secure and efficient energy exchanging in the energy network. In other words, the combined blockchain technology and edge computing is a promising platform for efficient and secure energy sharing in the energy market [91]. In order to overcome the unacceptable delay caused by blockchain technology in the energy sharing process, the edge-computing-based energy trading mechanism is an appropriate option for ensuring quick, secure, and efficient energy transactions in future modern energy grids.

3.4.5 Game theoretic-based energy trading mechanism

The game theory-based energy exchanging strategy is typically used for developing energy trading models that enable market participants for energy interactions under game theory techniques. This mechanism is more suitable for members who participate in competitive energy environments and looking for high-quality and affordable energy interactions.

In addition to the aforementioned energy trading mechanisms, there are also other mechanisms that are useful for developing effective energy exchanging models in line with the objectives of the related problem. These mechanisms are listed as follows and their descriptions can be fully found in [84].

- Real-time energy trading
- Peer-to-peer energy trading
- Blockchain-based energy trading
- Edge computing-based energy trading

- Game theoretic-based energy trading
- Bargaining-based cooperative energy trading
- Bidirectional energy trading
- Cross-border energy trading
- Emission-aware energy trading
- Event-driven energy trading
- Auction-based energy trading
- Machine learning-based energy trading
- Zero-energy trading
- Transactive energy trading

3.5 Summary

By attending to the importance of flexibility and sustainability in renewable-dominant energy grids, this chapter takes attention to the applications of energy trading strategies for manipulating practical solutions for future multi-carrier energy networks. Given the popularity and applicability of energy trading models in addressing the challenges of energy grids, their applications in recent literature were widely examined from both the operation and planning viewpoints. In this respect, the effectiveness of energy trading models was highlighted to indicate their properness for the renewable-based frameworks of future modern multi-energy grids. The necessity for going toward renewable-dominant energy grids was investigated to clearly specify the opportunities that they can create for energy grids given their feasibility, affordability, and viability in the energy network's plans. On the other hand, their inherent challenges were also discussed to analyze energy trading-based solutions for solving them. Due to this, energy trading mechanisms were suggested as one of the promising remedies for the uncontrollable property of RERs that can enable different energy network players for energy interactions in the hybrid energy infrastructure. For this aim, typical energy trading benefits were discussed in the first step. Then, different energy-sharing mechanisms were described to provide useful information regarding their attributes in energy transactions.

3.6 Questions

The careful analyzing whole of this chapter is expected to enable readers to effectively evaluate the following questions.

1) On which basis energy trading models are designed for energy network challenges?
2) What are the key roles of energy trading models for operation studies?

3) Which benefits can be reached by applying energy trading models to the energy operation problems?
4) Energy trading models are usually applicable for which types of operation problems?
5) How energy trading models can support energy planning problems in addressing the related challenges?
6) From the planning viewpoint, how energy trading models can be applicable for modern energy networks?
7) How can energy trading models affect the planning problems in energy management?
8) What are the reasons for the widespread usage of energy trading models in the energy operation and planning problems?
9) Why future energy networks are targeted to be a renewable-dominant structure of energy?
10) What are the motivations for going towards 100 percent RERs networks?
11) Which key opportunities and challenges can appear when the system operated under 100 percent RERs?
12) What are the key factors for the feasibility and economic viability of RERs and how do these factors affect the energy system's efficiency?
13) What is the recent progress in the implementation of 100 percent RERs plans around the world?
14) Why energy trading mechanisms can be appropriate solutions for the flexibility challenges of future renewable-dominant multi-energy networks?
15) How energy trading mechanisms can open new doors of flexibility for the sustainability of modern energy networks?
16) What are the general benefits of energy trading-based solutions for energy grids with a high or full penetration of RERs?
17) What are the different types of energy trading mechanisms and how they can be applied to various energy problems?
18) What are the diverse features of energy trading mechanisms applicable for future modern multi-carrier energy networks?

3.7 Suggestions

Given the topic of this chapter, the following suggestions are provided for the readers assisting them to pursue related research clues.

1) Developing adaptive energy trading models can facilitate the implementation possibility of grid modernization plans for future multi-carrier energy networks.

2) Analyzing the applicability of energy trading models from different viewpoints such as operation and planning can be led to obtaining a thorough overview for designing new energy sharing models in line with future energy network requirements.
3) Investigating the challenges and opportunities of renewable-dominant energy networks can drive the related plans to be effectively polished in accordance with energy network developments.
4) Designing capable energy sharing models for defending the realization of 100 percent RERs schemes can accelerate the successful switching from traditional energy structures to renewable-dominant modern energy networks.
5) Evaluating multifarious energy exchanging mechanisms can provide advancement opportunities for them to be suitable for the new structure of energy networks.

3.8 Future trends

Future trends can give a proper overview for the readers to form their related works and develop effective solutions for remaining gaps, which are listed as follows for this chapter.

1) Energy trading models need to be developed in a way to be adaptive to the implementation criteria of grid modernization plans for future modern multi-vector energy grids.
2) The assessment of the applicability of energy-sharing models is the necessary step for producing new energy transaction models that meet the related requirements of future energy grids.
3) Flexibility and sustainability are two key factors of future renewable-dominant energy grids that satisfying them at an acceptable level is essential for energy network developments.
4) Reaching 100 percent RERs goal is critical for future energy networks that need for advancing capable energy-sharing models to be feasible in practice.
5) Developing multi-energy trading mechanisms is an urgent action in response to ensuring the reliability of multi-energy supply in the system with a large number of stochastic producers.

References

[1] M. Vujanović, Q. Wang, M. Mohsen, N. Duić, J. Yan, Recent Progress in Sustainable Energy-Efficient Technologies and Environmental Impacts On Energy Systems, 283, Elsevier, 2021.
[2] M. Daneshvar, E. Babaei, Exchange market algorithm for multiple DG placement and sizing in a radial distribution system, Journal of Energy Management and Technology 2 (1) (2018) 54–65.

[3] B.P. Heard, B.W. Brook, T.M. Wigley, C.J. Bradshaw, Burden of proof: a comprehensive review of the feasibility of 100% renewable-electricity systems, Renew. Sust. Energ. Rev. 76 (2017) 1122–1133.
[4] D. Kirli, et al., Smart contracts in energy systems: a systematic review of fundamental approaches and implementations, Renew. Sust. Energ. Rev. 158 (2022) 112013.
[5] M. Daneshvar, S. Asadi, B. Mohammadi-Ivatloo, Technical and Theoretical Analysis of the Future Energy Network Modernization from Various Aspects, Grid Modernization—Future Energy Network Infrastructure, Springer, 2021, pp. 61–116.
[6] R. Shan, J. Reagan, S. Castellanos, S. Kurtz, N. Kittner, Evaluating emerging long-duration energy storage technologies, Renew. Sust. Energ. Rev. 159 (2022) 112240.
[7] M. Daneshvar, B. Mohammadi-ivatloo, K. Zare, Integration of distributed energy resources under the transactive energy structure in the future smart distribution networks, Operation of Distributed Energy Resources in Smart Distribution Networks, Elsevier, 2018, pp. 349–379.
[8] C. Choe, S. Cheon, J. Gu, H. Lim, Critical aspect of renewable syngas production for power-to-fuel via solid oxide electrolysis: integrative assessment for potential renewable energy source, Renew. Sust. Energ. Rev. 161 (2022) 112398.
[9] A. E. M. Operator, "And Energy Networks Australia (ENA), 'Open Energy Networks,' AEMO and ENA, Melbourne, VIC," Australia, Tech. Rep. 2018.
[10] M. Daneshvar, B. Mohammadi-Ivatloo, K. Zare, M. Abapour, S. Asadi, A. Anvari-Moghaddam, Chance-constrained scheduling of hybrid microgrids under transactive energy control, Int. J. Energy Res. 45 (7) (2021) 10173–10190.
[11] T. Wang, Y. Gong, C. Jiang, A review on promoting share of renewable energy by green-trading mechanisms in power system, Renew. Sust. Energ. Rev. 40 (2014) 923–929.
[12] Y. Li, C. Yu, Y. Liu, Z. Ni, L. Ge, X. Li, Collaborative Operation Between Power Network and Hydrogen Fueling Stations with Peer-to-Peer Energy Trading, IEEE Transactions on Transportation Electrification (2022).
[13] L. Wang, et al., Security constrained decentralized peer-to-peer transactive energy trading in distribution systems, CSEE Journal of Power and Energy Systems 8 (1) (2021) 188–197.
[14] W.-T. Lin, G. Chen, X. Zhou, Distributed carbon-aware energy trading of virtual power plant under denial of service attacks: a passivity-based neurodynamic approach, Energy (2022) 124751.
[15] S. Rezaei, A. Ghasemi, Stochastic scheduling of resilient interconnected energy hubs considering peer-to-peer energy trading and energy storages, J. Energy Storage 50 (2022) 104665.
[16] L.P.M.I. Sampath, A. Paudel, H.D. Nguyen, E.Y. Foo, H.B. Gooi, Peer-to-peer energy trading enabled optimal decentralized operation of smart distribution grids, IEEE Trans. Smart Grid 13 (1) (2021) 654–666.
[17] G. Li, Q. Li, X. Yang, R. Ding, General Nash bargaining based direct P2P energy trading among prosumers under multiple uncertainties, Int. J. Electr. Power Energy Syst. 143 (2022) 108403.
[18] J. Li, M.E. Khodayar, J. Wang, B. Zhou, Data-driven distributionally robust co-optimization of P2P energy trading and network operation for interconnected microgrids, IEEE Trans. Smart Grid 12 (6) (2021) 5172–5184.
[19] Y. Jia, C. Wan, P. Yu, Y. Song, P. Ju, Security Constrained P2P Energy Trading in Distribution Network: an Integrated Transaction and Operation Model, IEEE Trans. Smart Grid (2022).
[20] X. Zhao, L. Li, Y. Tao, S. Lai, X. Zhou, J. Qiu, Aggregated operation of heterogeneous small-capacity distributed energy resources in peer-to-peer energy trading, Int. J. Electr. Power Energy Syst. 141 (2022) 108162.

[21] C. Samende, J. Cao, Z. Fan, Multi-agent deep deterministic policy gradient algorithm for peer-to-peer energy trading considering distribution network constraints, Appl. Energy 317 (2022) 119123.
[22] Y. Jin, J. Choi, D. Won, Pricing and Operation Strategy for Peer-to-Peer Energy Trading Using Distribution System Usage Charge and Game Theoretic Model, IEEE Access 8 (2020) 137720–137730.
[23] M. Daneshvar, B. Mohammadi-Ivatloo, K. Zare, A novel transactive energy trading model for modernizing energy hubs in the coupled heat and electricity network, J. Clean Prod. 344 (2022) 131024.
[24] T. Huang, Y. Sun, M. Jiao, Z. Liu, J. Hao, Bilateral energy-trading model with hierarchical personalized pricing in a prosumer community, Int. J. Electr. Power Energy Syst. 141 (2022) 108179.
[25] M. Yunshou, W. Jiekang, C. Zhihong, W. Ruidong, Z. Ran, C. Lingmin, Cooperative operation framework for a wind-solar-CCHP multi-energy system based on Nash bargaining solution, IEEE Access 9 (2021) 119987–120000.
[26] M. Daneshvar, B. Mohammadi-Ivatloo, K. Zare, A. Anvari-Moghaddam, Transactive energy strategy for energy trading of proactive distribution company with renewable systems: a robust/stochastic hybrid technique, E-Prime-Advances in Electrical Engineering, 2, *Electronics and Energy*, 2022.
[27] C. Feng, B. Liang, Z. Li, W. Liu, F. Wen, Peer-to-Peer Energy Trading under Network Constraints Based on Generalized Fast Dual Ascent, IEEE Trans. Smart Grid (2022).
[28] X. Chang, Y. Xu, H. Sun, Vertex scenario-based robust peer-to-peer transactive energy trading in distribution networks, Int. J. Electr. Power Energy Syst. 138 (2022) 107903.
[29] M. Daneshvar, et al., Chance-constrained models for transactive energy management of interconnected microgrid clusters, J. Clean Prod. 271 (2020) 122177.
[30] M. Daneshvar, M. Pesaran, B. Mohammadi-ivatloo, Transactive energy integration in future smart rural network electrification, J. Clean Prod. 190 (2018) 645–654.
[31] M. Daneshvar, B. Mohammadi-Ivatloo, M. Abapour, S. Asadi, Energy exchange control in multiple microgrids with transactive energy management, J. Mod. Power Syst. Clean Energy 8 (4) (2020) 719–726.
[32] Y. Liu, L. Wu, J. Li, A two-stage peer-to-peer energy trading model for distribution systems with participation of utility, CSEE Journal of Power and Energy Systems 7 (5) (2021) 893–902.
[33] X. Luo, W. Shi, Y. Jiang, Y. Liu, J. Xia, Distributed peer-to-peer energy trading based on game theory in a community microgrid considering ownership complexity of distributed energy resources, J. Clean Prod. 351 (2022) 131573.
[34] W. Liu, D. Qi, F. Wen, Intraday residential demand response scheme based on peer-to-peer energy trading, IEEE Trans. Ind. Inf. 16 (3) (2019) 1823–1835.
[35] H. Görgülü, Y. Topçuoğlu, A. Yaldız, T. Gökçek, Y. Ateş, O. Erdinç, Peer-to-peer energy trading among smart homes considering responsive demand and interactive visual interface for monitoring, Sustainable Energy, Grids and Networks 29 (2022) 100584.
[36] W. Zhou, Y. Wang, F. Peng, Y. Liu, H. Sun, Y. Cong, Distribution network congestion management considering time sequence of peer-to-peer energy trading, Int. J. Electr. Power Energy Syst. 136 (2022) 107646.
[37] S. Aznavi, P. Fajri, M.B. Shadmand, A. Khoshkbar-Sadigh, Peer-to-peer operation strategy of PV equipped office buildings and charging stations considering electric vehicle energy pricing, IEEE Trans. Ind. Appl. 56 (5) (2020) 5848–5857.
[38] J. Guerrero, A.C. Chapman, G. Verbič, Decentralized P2P energy trading under network constraints in a low-voltage network, IEEE Trans. Smart Grid 10 (5) (2018) 5163–5173.
[39] M. Mehdinejad, H. Shayanfar, B. Mohammadi-Ivatloo, Peer-to-peer decentralized energy trading framework for retailers and prosumers, Appl. Energy 308 (2022) 118310.

[40] M. Daneshvar, B.M. Ivatloo, K. Zare, S. Asadi, A. Anvari-Moghaddam, A Stochastic Transactive Energy Model for Optimal Dispatch of Integrated Low-Carbon Energy Hubs in the Incorporated Electricity and Gas Networks, in: 2020 International Conference on Smart Grids and Energy Systems (SGES), IEEE, 2020, pp. 568–573.
[41] M. Daneshvar, B. Mohammadi-Ivatloo, M. Abapour, S. Asadi, R. Khanjani, Distributionally robust chance-constrained transactive energy framework for coupled electrical and gas microgrids, IEEE Trans. Ind. Electron. 68 (1) (2020) 347–357.
[42] M. Daneshvar, B. Mohammadi-Ivatloo, K. Zare, S. Asadi, Transactive energy management for optimal scheduling of interconnected microgrids with hydrogen energy storage, Int. J. Hydrogen Energy 46 (30) (2021) 16267–16278.
[43] Z. Wu, Z. Xu, W. Gu, S. Zhou, X. Yang, Decentralized Game-Based Robustly Planning Scheme for Distribution Network and Microgrids Considering Bilateral Energy Trading, IEEE Trans. Sustainable Energy 13 (2) (2021) 803–817.
[44] L. Ali, S. Muyeen, H. Bizhani, MG. Simoes, Economic Planning and Comparative Analysis of Market-driven Multi-microgrid system for Peer-to-Peer energy trading, IEEE Trans. Ind. Appl. 58 (3) (2022) 4025–4036.
[45] K.G.H. Kong, et al., Fuzzy optimization for P2P multi-period renewable energy trading planning, J. Clean Prod. (2022) 133122.
[46] L.P. Garcés, A.J. Conejo, R. García-Bertrand, R. Romero, A bilevel approach to transmission expansion planning within a market environment, IEEE Trans. Power Syst. 24 (3) (2009) 1513–1522.
[47] J.M. Delarestaghi, A. Arefi, G. Ledwich, A. Borghetti, A distribution network planning model considering neighborhood energy trading, Electric Power Systems Research 191 (2021) 106894.
[48] S. Favuzza, G. Graditi, M.G. Ippolito, E.R. Sanseverino, Optimal electrical distribution systems reinforcement planning using gas micro turbines by dynamic ant colony search algorithm, IEEE Trans. Power Syst. 22 (2) (2007) 580–587.
[49] A. Borghetti, et al., Impact of neighborhood energy trading and renewable energy communities on the operation and planning of distribution networks, Distributed Energy Resources in Local Integrated Energy Systems, Elsevier, 2021, pp. 125–174.
[50] A. Barnawi, S. Aggarwal, N. Kumar, D.M. Alghazzawi, B. Alzahrani, M. Boulares, Path Planning for Energy Management of Smart Maritime Electric Vehicles: a Blockchain-Based Solution, IEEE Trans. Intell. Transp. Syst. (2021).
[51] J. Liu, Y. Zhou, H. Yang, H. Wu, Uncertainty energy planning of net-zero energy communities with peer-to-peer energy trading and green vehicle storage considering climate changes by 2050 with machine learning methods, Appl. Energy 321 (2022) 119394.
[52] L. Ali, S. Muyeen, H. Bizhani, A. Ghosh, A multi-objective optimization for planning of networked microgrid using a game theory for peer-to-peer energy trading scheme, IET Generation, Transmission & Distribution 15 (24) (2021) 3423–3434.
[53] X. Chen, X. Wang, W. Ni, I.B. Collings, Two-way energy trading and online planning for fifth-generation communications with renewables, in: 2017 23rd Asia-Pacific Conference on Communications (APCC), IEEE, 2017, pp. 1–6.
[54] N.K. Meena, J. Yang, E. Zacharis, Optimal planning and operational management of open-market community microgrids, Energy Procedia 159 (2019) 533–538.
[55] M.R. Marinović, D.D. Makajić-Nikolić, M.J. Stanojević, Optimization in day-ahead planning of energy trading, Journal of applied engineering science 11 (4) (2013).
[56] J. Charkhonmattaku, C. Khompatraporn, Optimal VPP Energy Production Planning with Renewable Energy Sources and Small Dispatchable Generators, in: 2021 6th International Conference for Convergence in Technology (I2CT), IEEE, 2021, pp. 1–8.
[57] J. Dai, Y. Xiang, Will power company be profitable from non-utility-owned distributed generators planning with peer-to-peer strategy? Energy Reports 8 (2022) 147–154.

[58] H. Cong, X. Wang, C. Jiang, Two-stage nested bilevel model for generation expansion planning in combined electricity and gas markets, IET Generation, Transmission & Distribution 13 (15) (2019) 3443–3454.

[59] M.A. El-Sayed, Stochastic optimization of electric generation and trading-planning in liberalized market, in: 2009 IEEE/PES Power Systems Conference and Exposition, IEEE, 2009, pp. 1–7.

[60] J. Ramsebner, et al., From single to multi-energy and hybrid grids: historic growth and future vision, Renew. Sust. Energ. Rev. 151 (2021) 111520.

[61] M. Daneshvar, B. Mohammadi-ivatloo, S. Asadi, S. Galvani, Short term optimal hydro-thermal scheduling of the transmission system equipped with pumped storage in the competitive environment, Majlesi Journal of Electrical Engineering 14 (1) (2020) 77–84.

[62] M. Daneshvar, M. Abapour, B. Mohammadi-ivatloo, S. Asadi, Impact of optimal DG placement and sizing on power reliability and voltage profile of radial distribution networks, Majlesi Journal of Electrical Engineering 13 (2) (2019) 91–102.

[63] A.M. Adil, Y. Ko, Socio-technical evolution of Decentralized Energy Systems: a critical review and implications for urban planning and policy, Renew. Sust. Energ. Rev. 57 (2016) 1025–1037.

[64] M. Daneshvar, H. Eskandari, A.B. Sirous, R. Esmaeilzadeh, A novel techno-economic risk-averse strategy for optimal scheduling of renewable-based industrial microgrid, Sustainable Cities and Society 70 (2021) 102879.

[65] Z. Abba, N. Balta-Ozkan, P. Hart, A holistic risk management framework for renewable energy investments, Renew. Sust. Energ. Rev. 160 (2022) 112305.

[66] T.W. Brown, T. Bischof-Niemz, K. Blok, C. Breyer, H. Lund, B.V. Mathiesen, Response to 'Burden of proof: a comprehensive review of the feasibility of 100% renewable-electricity systems, Renew. Sust. Energ. Rev. 92 (2018) 834–847.

[67] N. Zhao, F. You, Sustainable power systems operations under renewable energy induced disjunctive uncertainties via machine learning-based robust optimization, Renew. Sust. Energ. Rev. 161 (2022) 112428.

[68] P. Arévalo, A. Cano, F. Jurado, Mitigation of carbon footprint with 100% renewable energy system by 2050: the case of Galapagos islands, Energy 245 (2022) 123247.

[69] C. Wu, X.-P. Zhang, M.J. Sterling, Economic Analysis of Power Grid Interconnections Among Europe, North-East Asia, and North America With 100% Renewable Energy Generation, IEEE Open Access Journal of Power and Energy 8 (2021) 268–280.

[70] F. Carere, F. Gatta, A. Geri, S. Lauria, M. Maccioni, and L. Nati, "Sardinia as a 100% Renewable Power System: a Frequency Stability Study (2022)."

[71] M. Daneshvar, B. Mohammadi-Ivatloo, K. Zare, S. Asadi, A. Anvari-Moghaddam, A novel operational model for interconnected microgrids participation in transactive energy market: a hybrid IGDT/stochastic approach, IEEE Trans. Ind. Inf. 17 (6) (2020) 4025–4035.

[72] T. Furubayashi, The role of biomass energy in a 100% renewable energy system for Akita prefecture, Japan, Energy Storage and Saving, 2022.

[73] S. Yin, J. Wang, Generation and transmission expansion planning towards a 100% renewable future, IEEE Trans. Power Syst. (2020).

[74] M. Daneshvar, B. Mohammadi-Ivatloo, K. Zare, S. Asadi, Two-stage robust stochastic model scheduling for transactive energy based renewable microgrids, IEEE Trans. Ind. Inf. 16 (11) (2020) 6857–6867.

[75] H.M. Tróndheim, B.A. Niclasen, T. Nielsen, F.F. Da Silva, C.L. Bak, 100% Sustainable electricity in the faroe islands: expansion planning through economic optimization, IEEE Open Access Journal of Power and Energy 8 (2021) 23–34.

[76] M. Diesendorf, B. Elliston, The feasibility of 100% renewable electricity systems: a response to critics, Renew. Sust. Energ. Rev. 93 (2018) 318–330.

[77] B. Kroposki, et al., Achieving a 100% renewable grid: operating electric power systems with extremely high levels of variable renewable energy, IEEE Power Energ. Mag. 15 (2) (2017) 61–73.
[78] B. Elliston, M. Diesendorf, I. MacGill, Simulations of scenarios with 100% renewable electricity in the Australian National Electricity Market, Energy Policy 45 (2012) 606–613.
[79] L.C. Kunkel, H.L. Breetz, J.K. Abbott, 100% renewable electricity policies in US cities: a mixed methods analysis of adoption and implementation, Energy Policy 167 (2022) 113053.
[80] B.V. Mathiesen, et al., Smart Energy Systems for coherent 100% renewable energy and transport solutions, Appl. Energy 145 (2015) 139–154.
[81] M. Ram, A. Gulagi, A. Aghahosseini, D. Bogdanov, C. Breyer, Energy transition in megacities towards 100% renewable energy: a case for Delhi, Renew. Energy 195 (2022) 578–589.
[82] J. Zhong, M. Bollen, S. Rönnberg, Towards a 100% renewable energy electricity generation system in Sweden, Renew. Energy 171 (2021) 812–824.
[83] D. Suna, G. Totschnig, F. Schöniger, G. Resch, J. Spreitzhofer, T. Esterl, Assessment of Flexibility Needs and Options For a 100% Renewable Electricity System By 2030 in Austria, *Smart Energy*, 2022.
[84] M. Daneshvar, S. Asadi, B. Mohammadi-Ivatloo, Energy trading possibilities in the modern multi-carrier energy networks, Grid Modernization—Future Energy Network Infrastructure, Springer, 2021, pp. 175–214.
[85] B. Miao, et al., Day-ahead energy trading strategy of regional integrated energy system considering energy cascade utilization, IEEE Access 8 (2020) 138021–138035.
[86] S. Bahrami, M.H. Amini, A decentralized framework for real-time energy trading in distribution networks with load and generation uncertainty, arXiv preprint (2017) *arXiv:1705.02575.*
[87] C. Zhang, J. Wu, Y. Zhou, M. Cheng, C. Long, Peer-to-Peer energy trading in a Microgrid, Appl. Energy 220 (2018) 1–12.
[88] J. Hamari, M. Sjöklint, A. Ukkonen, The sharing economy: why people participate in collaborative consumption, J. Assoc. Inf. Sci. Technol. 67 (9) (2016) 2047–2059.
[89] C. Eid, P. Codani, Y. Perez, J. Reneses, R. Hakvoort, Managing electric flexibility from Distributed Energy Resources: a review of incentives for market design, Renew. Sust. Energ. Rev. 64 (2016) 237–247.
[90] C. Burger, A. Kuhlmann, P. Richard, J. Weinmann, Blockchain in the energy transition. A survey among decision-makers in the German energy industry, DENA German Energy Agency 60 (2016) 1–44.
[91] Z. Zhou, L. Tan, G. Xu, Blockchain and edge computing based vehicle-to-grid energy trading in energy internet, in: 2018 2nd IEEE conference on energy internet and energy system integration (EI2), IEEE, 2018, pp. 1–5.

CHAPTER

4

The role of peer-to-peer energy trading strategy in future modern energy networks

4.1 Introduction

Over recent decades, energy grids have witnessed a tremendous transformation from the residential sector to the network-level part. This transformation is mostly accompanied by substantial changes in the energy generation sector when the disadvantages of incumbent energy suppliers are highlighted especially in terms of economic and environmental [1]. In this regard, the idea of producing clean energy using the environmental potential has been flourished and created significant evolution in the energy generation part. This issue has been led to the attractive attention to renewable energy sources (RESs) and their environmental and economic benefits have made them popular units for power generation [2]. Following this popularity for RESs, their technology developments have resulted in the emergence of different types of renewable systems that enable the grid to effectively utilize the multifarious natural energy sources for multi-energy production. Hence, RESs have become one of the main sources of energy in the power system. Indeed, their impressive advantages have moved the most of energy generation sector's reliance from fossil fuel-based units to renewable systems [3]. However, this transition has been accompanied by critical challenges that directly back to the uncertain nature of RESs [4]. In other words, the power production by RESs directly depends on climate

Emerging Transactive Energy Technology for Future Modern Energy Networks
DOI: https://doi.org/10.1016/B978-0-323-91133-7.00005-3

changes and their uncertainties enter stochastic intermittences on RESs outputs [5]. The uncontrollable power production by RESs has challenged the reliable scheduling of the power grid and has increased the fluctuations in the energy generation sector [6]. Due to this, several approaches are proposed for diminishing the unfavorable effects of RESs' uncertainties. Herein, optimal integration of RESs and energy storage systems are recognized as promising ways for enhancing the reliability of the system in the presence of renewables [7]. However, some of these solutions such as energy storage systems can impose more costs for the system's operation. In this regard, energy trading techniques have appeared as one of the confident ways for increasing the flexibility of renewable-based systems [8]. This issue has been led to the development of new strategies of energy exchanging that each of them offers specific advantages for the system. One of these strategies is peer-to-peer (P2P) energy trading, which has attracted tremendous attention in recent years [9]. The P2P enables the grid to establish direct energy exchange among different network platforms in the local area. This energy sharing strategy can upsurge the effectiveness of grid exploitation [10]. The P2P energy trading can facilitate the promotion of local grids with a high share of renewables and substantially reduce the need for expanding or creating new energy generation plants and transmission lines with considerable power losses. This chapter is focused on investigating the role of the P2P energy trading strategy in future modern hybrid energy networks. In this respect, the applications of P2P energy trading in energy grids are scrutinized at the first step. Then, the chapter includes information regarding the P2P energy trading market and its attributes.

4.2 Peer-to-Peer energy trading applications

The P2P energy exchanging strategy has been used as one of the promising ways for increasing the effectiveness of the grid operation. This strategy is recently applied to different problems as an appropriate solution for the related challenges. For example, an equilibrium model of P2P energy sharing is proposed in [11] by deploying the shared energy storage in the end-user part. Aiming to achieve cost-saving and improve energy efficiency, the authors in [12] proposed a P2P energy trading framework considering multi-energy coupling for buildings. In [13], a P2P energy trading strategy is used for designing a new platform among residential homes with the aim of coordinating demand response programs as well as reducing energy disturbances in the system. A decentralized P2P energy sharing is applied in [14] for proposing a comprehensive energy trading framework intending multiple dynamic components in smart buildings with electric vehicles and battery storage systems. A novel and fair P2P energy sharing framework is presented in [15] to coordinately manage the energy of building with the aim of realizing economic and sustainable objectives. The authors in [16] proposed an individual P2P energy trading mechanism for the energy management of a net-zero community considering the allocation of an individual peer trading price. The P2P energy trading model is presented in [17] for developing an active market structure aiming st the energy management of household models. In [18], blockchain technology is used for facilitating P2P energy sharing for investigating system security, scalability, and decentralization. A privacy-preserving operation model of the P2P energy sharing model is proposed in [19] based on enhanced Benders decomposition to optimally schedule transacted energy between smart buildings. A transactive energy mechanism is used in [9] for developing a P2P energy trading model to create energy trading possibilities for modern energy hubs aiming to increase their flexibilities in the coupled heat and electricity grid. In [20], a P2P energy trading framework is proposed for the wind power producer to not only promote wind power usage but also increase its revenue in the retailing layer of the system. The application of P2P energy trading is not limited only to mentioned areas and it is exerted in several works that some of them are listed as follows.

- Optimal sizing and economic analysis of RESs [21];
- Developing a two-level hierarchical incentive mechanism [22];
- Least-cost energy path optimization for prosumers matching [23];
- Multi-commodity optimization of energy and communication resources in P2P energy sharing resources [24];
- The energy balance service provider [25];
- Optimal energy management of grid-connected prosumers [26];
- Electric vehicle P2P energy trading model [27];
- A secured advanced management architecture for multi-microgrid [28];
- Multiclass energy management for prosumers [29];
- Multi-agent microgrid management system [30];
- Energy sharing possibility among smart energy buildings [31];
- Optimal decentralized operation of smart distribution networks [32];
- Strategic energy market design for community microgrids [33];
- Reliability-oriented operation of distribution grids [34];
- Scalable coordinated management of P2P energy sharing [35];
- Coordinated energy market design in distribution networks [36];
- A two-tier distributed energy market clearing scheme [37];
- Energy management of grid integrated multi-microgrids [38];
- Optimal bidding strategy for ancillary service provision [39];
- P2P energy exchanging among smart homes [40];
- Security constrained P2P energy sharing in the distribution grid [41].

4.3 Peer-to-Peer energy trading market

The decentralization promotion is increasing day by day in response to the call for affordable decarbonization schemes [42]. From one key aspect, the effectiveness of the related schemes mostly relies on how to design the energy market [33]. This issue also mostly backs to the vital role of energy markets in the provision of energy trading opportunities for the market participants and allowing them to properly follow their various objectives. In this regard, energy markets are developed based on different strategies that one of the important of them is P2P energy trading. Until now, several P2P strategies are suggested for the P2P energy trading markets that typically are different in the topology and degree of decentralization and are [43,44]:

- **Full P2P market:** This type of energy trading market allows peers to directly negotiate with each other aiming to determine the energy sell or buy in the system. Thus, the amount of transacted energy among two peers and its price are determined without centralized supervision.
- **Community-based market:** The energy interactions in this market stand on the presence of a community manager who is responsible for controlling energy activities in the community. In this market, a community manager can play an intermediator role among the community and the rest of the structure.
- **Hybrid P2P market**: This energy sharing market structure is the combination of two previous ones that end up with multifarious layers of energy exchange.

The P2P energy trading market provides different advantages in its various structures. These advantages are illustrated in Fig. 4.1 [44,45]. The full P2P energy trading market realizes the full democratization in energy usage. Participants possess freedom of choice and autonomy when they interact under the full P2P energy market guidelines. Moreover, energy use aligned with agents' preferences in this structure. In the community-based market, P2P energy trading upsurges the relationships and involvement of community participants in the system. It mobilizes the social cooperation and resilience of the system's participants. Furthermore, the community manager procures opportunities for new services for grid operators. In another structure of the P2P energy sharing market, the hybrid P2P market offers a scalable energy interaction area for market members. This market structure can provide the most compatible conditions for the system and it is also more predictable for grid operators. Although P2P energy exchanging markets bring tremendous advantages for the system, their successful operation poses considerable challenges that need innovative solutions for

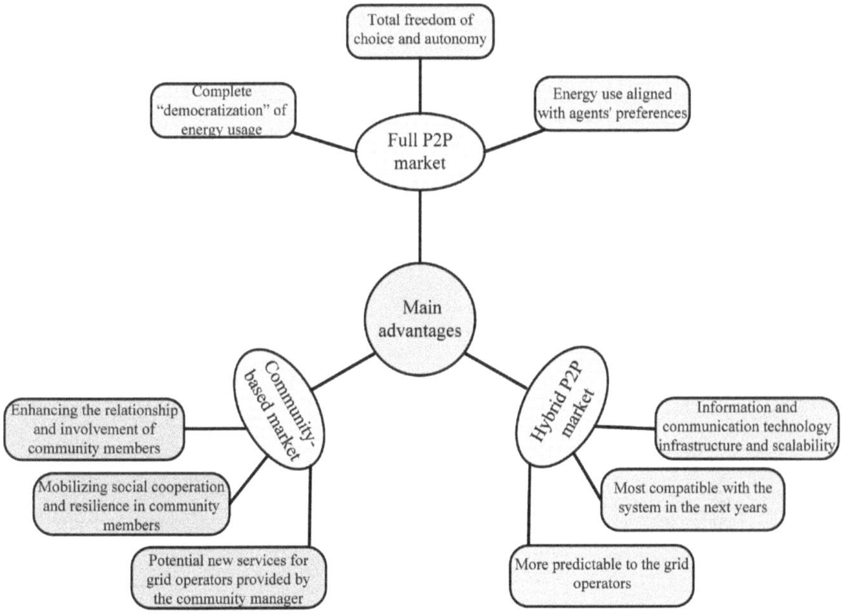

FIGURE 4.1 Main advantages of different P2P energy trading market structures.

addressing them. In this respect, the main challenges for the P2P energy trading markets are demonstrated in Fig. 4.2 [44,45].

Among different P2P markets, the full P2P energy trading market faces challenges regarding the investment, maintenance, and scalability of the system. It also requires to be effectively upgraded for providing high-quality and safe energy. This market needs to forecast the system behavior as it suffers from centralized control. Delaying on energy delivery can also be occurred under this market due to the slow convergence in achieving a consensus in the system. The community-based P2P market structure needs to be improved for properly managing preferences of energy usage. Privacy-preserving as well as procuring a fair and unbias energy exchanging among community members are other challenges in this structure. This is while coordinating internal trades in the system is the concern for the hybrid P2P energy trading market.

4.4 Peer-to-Peer energy attributes

The P2P energy trading market possesses different key attributes regarding its potential and effectiveness. Herein, better analyzing these attributes can enable the system to effectively use its potential in developing

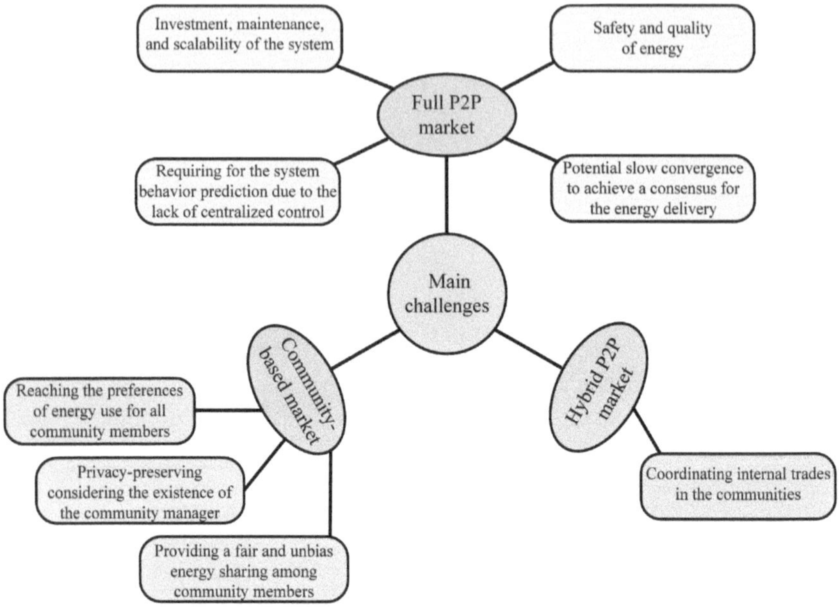

FIGURE 4.2 Main challenges of different P2P energy trading market structures.

future modern energy networks. Indeed, as future modern energy networks are targeted for 100 percent clean energy production, flexibility is one of the key factors that needs to be provided in different ways [46]. As energy trading strategies can allow the system to be more flexible in energy serving [47], the P2P energy trading mechanism can be a suitable way for satisfying the flexibility requirements of the system. The attributes of P2P energy trading include its strengths, weaknesses, opportunities, and threats are described in the following subsections [45].

4.4.1 P2P strengths

The P2P energy trading market is recognized as one of the famous methods for its strengths in facilitating the realization of the goals of modern energy grids. The different strengths of the P2P energy trading market structure are illustrated in Fig. 4.3 [45]. One of its strengths undoubtedly backs to its ability in improving consumers' choice as well as transparency. The P2P energy trading market is more flexible in providing diverse choices regarding the type and origin of the electricity. Thus, it develops a more collaboratively environment for energy interactions among market participants. For example, reference [48] presents five various classes of collaboration in P2P energy markets for consumers of an urban area. The reliability of energy supply is a critical issue for systems that face a high

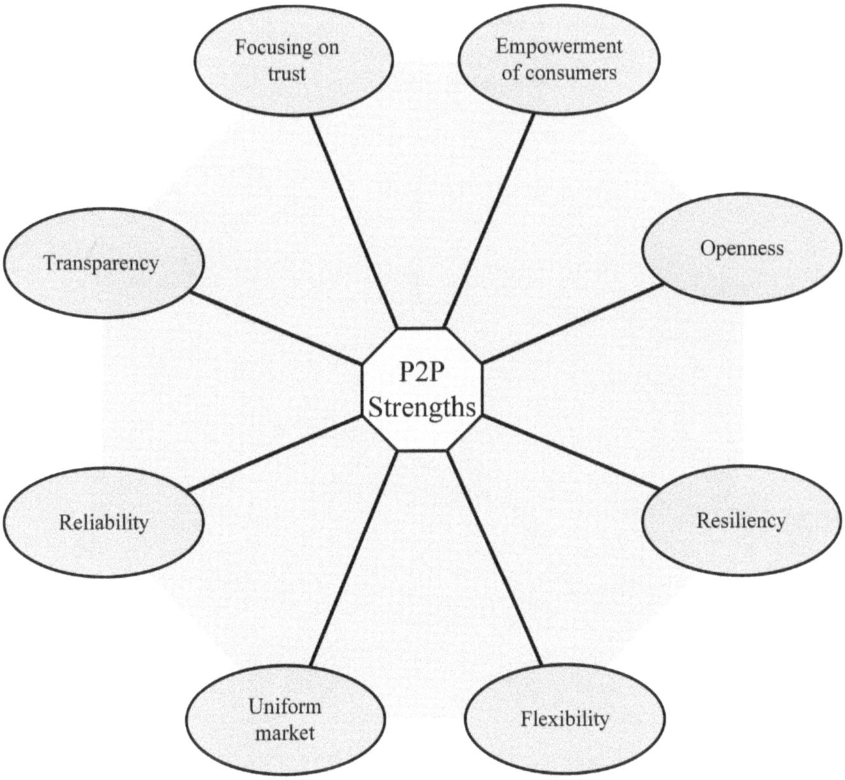

FIGURE 4.3 Strengths of the P2P energy trading market structure.

level of RESs. Due to uncertainties in RESs production, it is essential to adopt appropriate strategies for increasing the ways of energy supply in the presence of RESs [49]. The P2P energy trading enables the system to effectively establish energy trading possibilities among market members subsequently empowers the grid for utilizing energy trading as another promising way for increasing the reliability of energy supply in the system. The P2P markets offer a uniform process by removing the potential market power related to special players. They can improve the resiliency of the grid by making energy sharing possible for different structures of the grid. Another significant point of P2P energy trading markets stands on their ability in enhancing the flexibility of the network in the presence of RESs. Indeed, the system can use P2P energy trading as a proper strategy in serving the energy demand that increases the system's flexibility in balancing energy. All aforementioned strengths have converted P2P energy exchanging markets to one of the undeniable ways of realizing the goals of future modern energy grids and facilitating the grid modernization process.

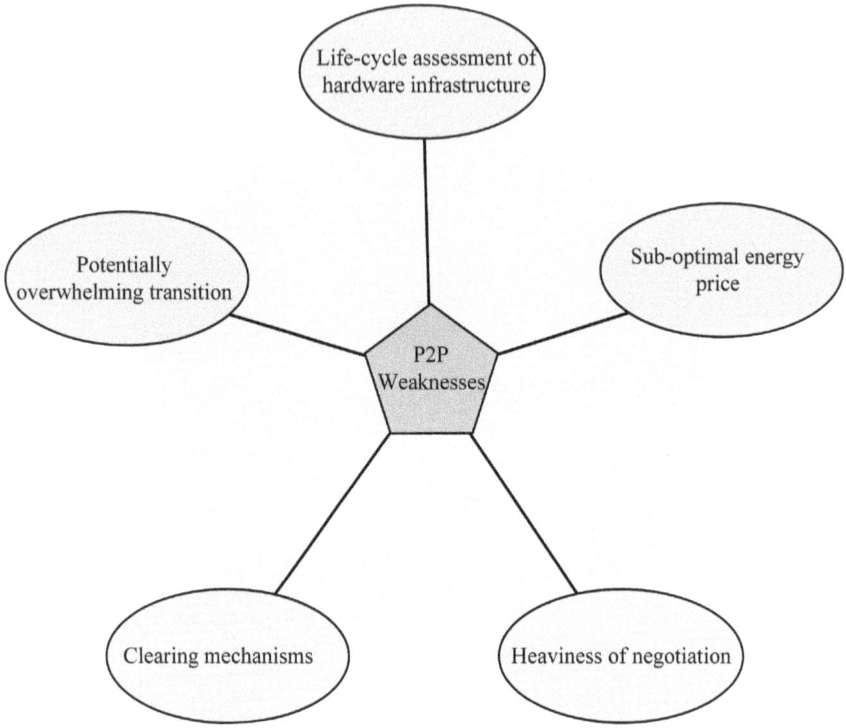

FIGURE 4.4 Weaknesses of the P2P energy trading market structure.

4.4.2 P2P weaknesses

In versus significant strengths for P2P energy trading markets, they possess substantial weaknesses in playing their key role that are need to be assessed for effectively addressing in the body of the grid. Fig. 4.4 demonstrates various weaknesses of the P2P energy trading market structure [45]. In the energy interactions of P2P energy markets, the market design needs to consider consumers' preferences to make product differentiation possible for them. Intending preferences in energy trading leads to various prices for each energy transaction. This process can be led to a sub-optimization of the overall price that makes the process complex and highlights one of the obvious weaknesses of P2P markets [45]. This is while simultaneously negotiating all market participants with all others can ease the convergence of prices to similar values [45]. In the P2P market structure, as a large number of participants are involved in energy interactions, the system is expected to face a high level of energy transactions. As another weakness, the huge volume of energy transactions indicates the need for advanced monitoring and managing system to make

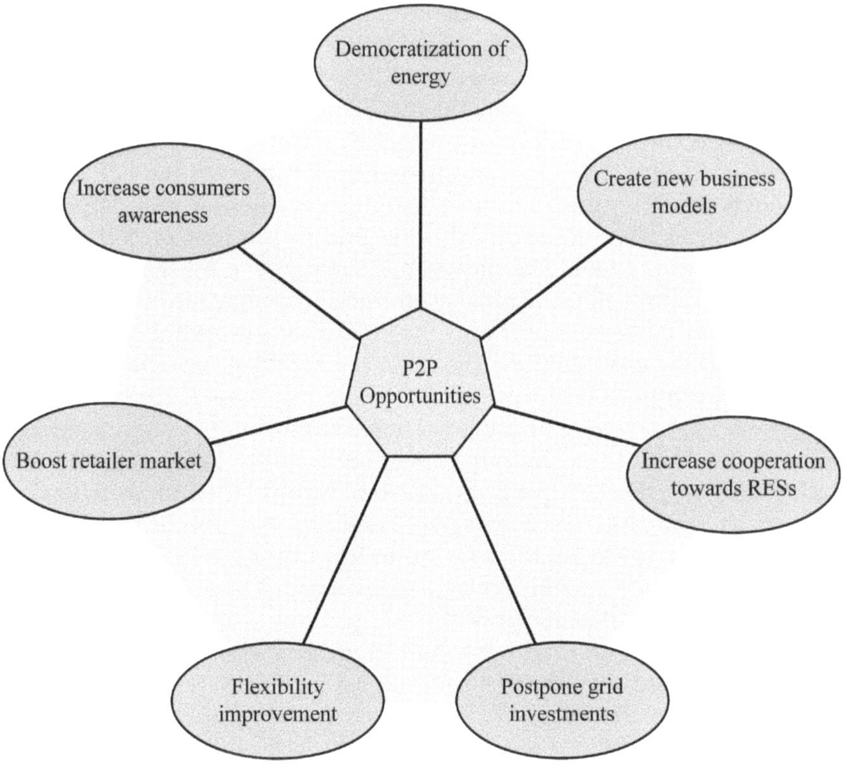

FIGURE 4.5 Opportunities of the P2P energy trading market structure.

market objectives feasible [50]. In addition to the overwhelming number of energy transactions, the negotiation mechanism is another effective factor in realizing the successful P2P energy trading paradigm making serious remedies necessary to be suitable for the future modern energy network infrastructure.

4.4.3 P2P opportunities

As mentioned, P2P energy trading markets can support the system for a high-level deployment of RESs as alternative energy production systems for conventional power plants with fossil fuels. Despite the existence of different strengths and weaknesses for P2P energy markets, they can bring tremendous opportunities for overcoming challenges that already were barriers to the efficient operation of the system. Some of the important opportunities of the P2P energy trading market structure are illustrated in Fig. 4.5 [45]. One of these opportunities is the democratization of energy that can be reached under the P2P energy market paradigm. The creation of this opportunity mostly backs to the ability of the P2P-based system for procuring energy interactions by considering participants' preferences as well as suggesting diverse energy products. This type of market enables participants to benefit from new business models that are created to provide cost-effective conditions for energy interactions of market members. Boost retail market is another prominent opportunity of the P2P market that allows retail companies to have energy transactions in the competitive environment as they could not do this due to the lack of competition in many countries [45]. This is while P2P energy markets allow retail companies to adopt effective business models for their energy transactions in the market environment. From the economic viewpoint, P2P energy trading markets can postpone the need for building new power plants or transmission lines with expensive equipment. Indeed, the established energy trading allows participants to support each other by sharing energy during a day without exclusively depending on central energy generation units [51]. Thus, market operators benefit from this structure by deferring network investments in energy units, lines, and equipment. Finally, P2P energy trading can increase cooperation towards renewable systems. In other words, future modern energy grids are targeted for a fully clean energy production that requires capable strategies for properly dealing with intermittences in their outputs. Herein, P2P energy trading can be one of the promising mechanisms for improving the system's flexibility in capturing the uncertainties of RESs production and reliably serving energy for various end-users.

4.4.4 P2P threats

In versus opportunities for the P2P energy market, the system can be faced with remarkable threats when it operates under the P2P energy markets paradigm. Fig. 4.6. portrays several key threats of the P2P energy trading market structure [45]. One of these threats can be a legal framework and the related limitations in different countries as it is expected to be removed by more clarifying the advantages of P2P markets, especially their role in facilitating the adoption of renewable systems. On the other hand, communities with less economic power or energy poverty of some group of consumers can appear when the grid is empowered by P2P markets implementation. Herein, another point of concern stands on the willingness of prosumers and consumers for participating in P2P energy trading interactions and their engagement in the market area. In addition to the mentioned threats, technology dependency for successfully implementing P2P markets, security of the system, and privacy of the gathered data are other key issues that need to be effectively addressed in the P2P energy trading environment.

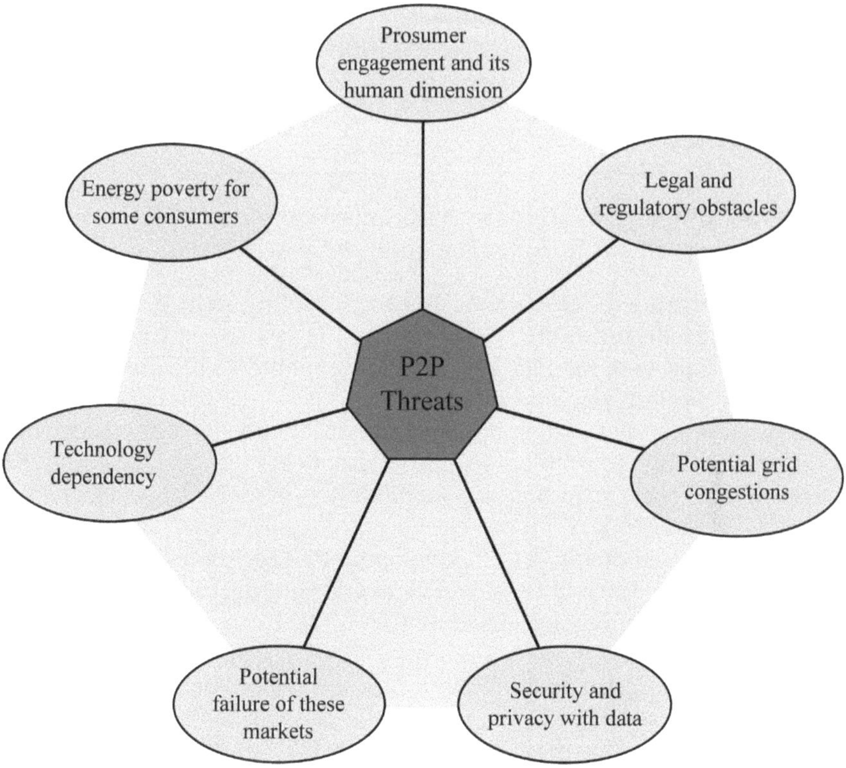

FIGURE 4.6 Threats of the P2P energy trading market structure.

4.5 Summary

In response to the need for practical solutions for supporting energy grids towards modernization, this chapter analyzed the role of P2P energy trading in modernizing future energy grids. In order to highlight the widespread effectiveness of P2P energy trading in energy problems, its application was investigated in several recent studies with different problems. The full P2P energy trading market, community-based market, and hybrid P2P market were evaluated as different types of the P2P market. Moreover, their advantages and challenges were described to provide effective guidance regarding their expected presence in modern energy networks' interactions. Generally, familiarizing with the properties of a certain issue can provide a useful overview of how its usage in the related field. Due to this, important attributes of P2P energy trading markets were explained aiming to give an appropriate overview for their cooperation in future modern energy networks. In this respect, strengths, weaknesses, opportunities, and threats as four key attributes of P2P energy markets were described and their effects on energy interactions were also discussed in detail.

4.6 Questions

The careful analyzing whole of this chapter is expected to enable readers to effectively evaluate the following questions.

1) Why analyzing the role of the P2P energy trading method is important for grid modernization?
2) Which aspects of the P2P energy trading method have made it a key tool for modernizing energy grids?
3) What are the distinguish applications of P2P energy trading mechanisms in different problems of the energy field?
4) How P2P energy trading can facilitate the modernization of energy networks?
5) What are the motivations for developing P2P energy trading markets?
6) What are the different types of P2P energy sharing markets?
7) From which viewpoints the full P2P market can be useful for grid modernization?
8) Which challenges can be appeared in the system under the full P2P market operation?
9) What are the motivations for developing P2P community-based markets?
10) From which aspects the P2P community-based market can be challengeable for the system?
11) What are the objectives for creating hybrid P2P markets?
12) The system can be faced with which advantages and challenges when it works under the hybrid P2P market paradigm?
13) Why investigating key attributes of P2P energy trading markets is essential for effectively modernizing future energy networks?
14) What are the key properties of P2P energy trading markets in the energy grid infrastructure?
15) What are the strengths of the P2P energy trading market in the energy grid?
16) Which weaknesses of the P2P energy sharing market can be problematic in achieving systems' goals?
17) Which opportunities can be reached from the exploitation of the system under the P2P energy sharing paradigm?
18) The grid can be faced with which threats when it establishes energy interactions based on the P2P energy sharing market conditions?
19) How P2P energy trading markets can improve the systems' flexibility, reliability of supply, and sustainability in modern energy grids?

4.7 Suggestions

Given the topic of this chapter, the following suggestions are provided for the readers assisting them to pursue related research clues.

1) Analyzing the critical role of the P2P energy trading strategy can facilitate its effective usage in the grid modernization process.
2) Scrutinizing different applications of P2P energy trading mechanisms in energy studies can effectively highlight its applicability for future modern energy networks.
3) Developing multifarious types of P2P energy trading markets can enable the system to make a holistic structure adaptable to the goals of modern energy grids.
4) Investigating key attributes of P2P energy sharing markets can provide useful information regarding their potential for deploying in the modern energy network infrastructure.
5) Analyzing the effects of P2P energy trading implementation on energy interactions in the body of the grid can empower the system for reaching to alternative complete structure.

4.8 Future trends

Future trends can give a proper overview for the readers to form their related works and develop effective solutions for remaining gaps, which are listed as follows for this chapter.

1) The crucial role of the P2P energy trading strategy needs to be investigated in detail aiming to clearly highlight its potential for applying to the modern energy network infrastructure from different dimensions.
2) The applicability of the P2P energy trading strategy is a key factor that exploring it is essential for achieving the scheduled goals of grid modernization.
3) A holistic P2P energy trading market structure needs to be developed in a way to intend the involvement of various market participants with different playing patterns.
4) Evaluating diverse attributes of P2P energy trading markets is necessary to give a comprehensive overview of its capability for employing in energy interactions of future modern energy networks.
5) All effects of P2P energy trading implementation need to be examined with the aim of creating suitable development opportunities for energy trading structures.

References

[1] M. Daneshvar, E. Babaei, Exchange market algorithm for multiple DG placement and sizing in a radial distribution system, Journal of Energy Management and Technology 2 (1) (2018) 54–65.
[2] D. Abdul, J. Wenqi, A. Tanveer, Prioritization of renewable energy source for electricity generation through AHP-VIKOR integrated methodology, Renew. Energy 184 (2022) 1018–1032.
[3] A.A. Kebede, T. Kalogiannis, J. Van Mierlo, M. Berecibar, A comprehensive review of stationary energy storage devices for large scale renewable energy sources grid integration, Renew. Sust. Energ. Rev. 159 (2022) 112213.
[4] H. Karimi, S. Jadid, Modeling of transactive energy in multi-microgrid systems by hybrid of competitive-cooperative games, Electric Power Systems Research 201 (2021) 107546.
[5] M. Daneshvar, B. Mohammadi-ivatloo, S. Asadi, S. Galvani, Short term optimal hydro-thermal scheduling of the transmission system equipped with pumped storage in the competitive environment, Majlesi Journal of Electrical Engineering 14 (1) (2020) 77–84.
[6] H. Karimi, S. Jadid, A. Makui, Stochastic energy scheduling of multi-microgrid systems considering independence performance index and energy storage systems, J. Energy Storage 33 (2021) 102083.
[7] M. Daneshvar, H. Eskandari, A.B. Sirous, R. Esmaeilzadeh, A novel techno-economic risk-averse strategy for optimal scheduling of renewable-based industrial microgrid, Sustainable Cities and Society 70 (2021) 102879.
[8] W. Tushar, et al., A coalition formation game framework for peer-to-peer energy trading, Appl. Energy 261 (2020) 114436.
[9] M. Daneshvar, B. Mohammadi-Ivatloo, K. Zare, A novel transactive energy trading model for modernizing energy hubs in the coupled heat and electricity network, J. Clean Prod. (2022) 131024.
[10] Y.-C. Tsao, V-V. Thanh, Toward sustainable microgrids with blockchain technology-based peer-to-peer energy trading mechanism: a fuzzy meta-heuristic approach, Renew. Sust. Energ. Rev. 136 (2021) 110452.
[11] B. Zheng, W. Wei, Y. Chen, Q. Wu, S. Mei, A peer-to-peer energy trading market embedded with residential shared energy storage units, Appl. Energy 308 (2022) 118400.
[12] H. Zhang, S. Zhang, X. Hu, H. Cheng, Q. Gu, M. Du, Parametric optimization-based peer-to-peer energy trading among commercial buildings considering multiple energy conversion, Appl. Energy 306 (2022) 118040.
[13] W. Liu, D. Qi, F. Wen, Intraday residential demand response scheme based on peer-to-peer energy trading, IEEE Trans. Ind. Inf. 16 (3) (2019) 1823–1835.
[14] C. Lyu, Y. Jia, Z. Xu, Fully decentralized peer-to-peer energy sharing framework for smart buildings with local battery system and aggregated electric vehicles, Appl. Energy 299 (2021) 117243.
[15] S. Cui, Y.-W. Wang, Y. Shi, J.-W. Xiao, A new and fair peer-to-peer energy sharing framework for energy buildings, IEEE Trans. Smart Grid 11 (5) (2020) 3817–3826.
[16] J. Liu, H. Yang, Y. Zhou, Peer-to-peer energy trading of net-zero energy communities with renewable energy systems integrating hydrogen vehicle storage, Appl. Energy 298 (2021) 117206.
[17] H. Görgülü, Y. Topçuoğlu, A. Yaldız, T. Gökçek, Y. Ateş, O. Erdinç, Peer-to-peer energy trading among smart homes considering responsive demand and interactive visual interface for monitoring, Sustainable Energy, Grids and Networks 29 (2022) 100584.
[18] P. Wongthongtham, D. Marrable, B. Abu-Salih, X. Liu, G. Morrison, Blockchain-enabled Peer-to-Peer energy trading, Computers & Electrical Engineering 94 (2021) 107299.

[19] Y. Xia, Q. Xu, S. Tao, P. Du, Y. Ding, J. Fang, Preserving operation privacy of peer-to-peer energy transaction based on Enhanced Benders Decomposition considering uncertainty of renewable energy generations, Energy (2022) 123567.

[20] H. Rashidizadeh-Kermani, M. Vahedipour-Dahraie, M. Shafie-khah, P. Siano, A Peer-to-Peer Energy Trading Framework for Wind Power Producers With Load Serving Entities in Retailing Layer, IEEE Syst. J. (2021).

[21] A. Yaldız, T. Gökçek, İ. Şengör, O. Erdinç, Optimal sizing and economic analysis of Photovoltaic distributed generation with Battery Energy Storage System considering peer-to-peer energy trading, Sustainable Energy, Grids and Networks 28 (2021) 100540.

[22] Y. Wang, et al., Modelling and analysis of a two-level incentive mechanism based peer-to-peer energy sharing community, Int. J. Electr. Power Energy Syst. 133 (2021) 107202.

[23] O. Jogunola, W. Wang, B. Adebisi, Prosumers matching and least-cost energy path optimisation for peer-to-peer energy trading, IEEE Access 8 (2020) 95266–95277.

[24] O. Jogunola, B. Adebisi, K. Anoh, A. Ikpehai, M. Hammoudeh, G. Harris, Multi-commodity optimisation of peer-to-peer energy trading resources in smart grid, Journal of Modern Power Systems and Clean Energy (2021).

[25] Z. Wang, X. Yu, Y. Mu, H. Jia, Q. Jiang, X. Wang, Peer-to-Peer energy trading strategy for energy balance service provider (EBSP) considering market elasticity in community microgrid, Appl. Energy 303 (2021) 117596.

[26] K. Kusakana, Optimal peer-to-peer energy management between grid-connected prosumers with battery storage and photovoltaic systems, J. Energy Storage 32 (2020) 101717.

[27] Z. Li, S. Chen, B. Zhou, Electric vehicle peer-to-peer energy trading model based on smes and blockchain, IEEE Trans. Appl. Supercond. 31 (8) (2021) 1–4.

[28] M.A. Mohamed, A. Hajjiah, K.A. Alnowibet, A.F. Alrasheedi, E.M. Awwad, S. Muyeen, A secured advanced management architecture in peer-to-peer energy trading for multi-microgrid in the stochastic environment, IEEE Access 9 (2021) 92083–92100.

[29] T. Morstyn, M.D. McCulloch, Multiclass energy management for peer-to-peer energy trading driven by prosumer preferences, IEEE Trans. Power Syst. 34 (5) (2018) 4005–4014.

[30] L. Gomes, Z.A. Vale, J.M. Corchado, Multi-agent microgrid management system for single-board computers: a case study on peer-to-peer energy trading, IEEE Access 8 (2020) 64169–64183.

[31] S. Cui, Y.-W. Wang, J.-W. Xiao, Peer-to-peer energy sharing among smart energy buildings by distributed transaction, IEEE Trans. Smart Grid 10 (6) (2019) 6491–6501.

[32] L.P.M.I. Sampath, A. Paudel, H.D. Nguyen, E.Y. Foo, H.B. Gooi, Peer-to-Peer Energy Trading Enabled Optimal Decentralized Operation of Smart Distribution Grids, IEEE Trans. Smart Grid 13 (1) (2021) 654–666.

[33] Z. Zhang, H. Tang, J. Ren, Q. Huang, W.-J. Lee, Strategic prosumers-based peer-to-peer energy market design for community microgrids, IEEE Trans. Ind. Appl. 57 (3) (2021) 2048–2057.

[34] P. Bayat, H. Afrakhte, P. Bayat, Reliability-oriented operation of distribution networks with multi-microgrids considering peer-to-peer energy sharing, Sustainable Energy, Grids and Networks 28 (2021) 100530.

[35] D. Qiu, Y. Ye, D. Papadaskalopoulos, G. Strbac, Scalable coordinated management of peer-to-peer energy trading: a multi-cluster deep reinforcement learning approach, Appl. Energy 292 (2021) 116940.

[36] K. Zhang, S. Troitzsch, S. Hanif, T. Hamacher, Coordinated market design for peer-to-peer energy trade and ancillary services in distribution grids, IEEE Trans. Smart Grid 11 (4) (2020) 2929–2941.

[37] M.H. Ullah, J.-D. Park, A two-tier distributed market clearing scheme for peer-to-peer energy sharing in smart grid, IEEE Trans. Ind. Inf. 18 (1) (2021) 66–76.

[38] K. Thirugnanam, M.S. El Moursi, V. Khadkikar, H.H. Zeineldin, M. Al Hosani, Energy Management of Grid Interconnected Multi-Microgrids Based on P2P Energy Exchange: a Data Driven Approach, IEEE Trans. Power Syst. 36 (2) (2020) 1546–1562.
[39] Y. Zhou, J. Wu, G. Song, C. Long, Framework design and optimal bidding strategy for ancillary service provision from a peer-to-peer energy trading community, Appl. Energy 278 (2020) 115671.
[40] M.R. Alam, M. St-Hilaire, T. Kunz, Peer-to-peer energy trading among smart homes, Appl. Energy 238 (2019) 1434–1443.
[41] Y. Jia, C. Wan, P. Yu, Y. Song, P. Ju, Security Constrained P2P Energy Trading in Distribution Network: an Integrated Transaction and Operation Model, IEEE Trans. Smart Grid (2022).
[42] M. Daneshvar, B. Mohammadi-Ivatloo, K. Zare, M. Abapour, S. Asadi, A. Anvari-Moghaddam, Chance-constrained scheduling of hybrid microgrids under transactive energy control, Int. J. Energy Res. 45 (7) (2021) 10173–10190.
[43] H. Beitollahi, G. Deconinck, Peer-to-peer networks applied to power grid, in: Proceedings of the International conference on Risks and Security of Internet and Systems (CRiSIS), 8, Citeseer, 2007.
[44] Y. Parag, B. Sovacool, Electricity market design for the prosumer era, Nat Energy 1 (16) (2016) 032.
[45] T. Sousa, T. Soares, P. Pinson, F. Moret, T. Baroche, E. Sorin, Peer-to-peer and community-based markets: a comprehensive review, Renew. Sust. Energ. Rev. 104 (2019) 367–378.
[46] M. Daneshvar, B. Mohammadi-ivatloo, K. Zare, Integration of distributed energy resources under the transactive energy structure in the future smart distribution networks, in: Operation of Distributed Energy Resources in Smart Distribution Networks, Elsevier, 2018, pp. 349–379.
[47] M. Daneshvar, B. Mohammadi-Ivatloo, K. Zare, A. Anvari-Moghaddam, Transactive Energy Strategy for Energy Trading of Proactive Distribution Company with Renewable Systems: a Robust/Stochastic Hybrid Technique, e-Prime-Advances in Electrical Engineering, Electronics and Energy (2022) 100028.
[48] T. Brudermann, Y. Yamagata, Towards an agent-based model of urban electricity sharing, in: 2014 International Conference and Utility Exhibition on Green Energy for Sustainable Development (ICUE), IEEE, 2014, pp. 1–5.
[49] M. Daneshvar, B. Mohammadi-Ivatloo, K. Zare, S. Asadi, A. Anvari-Moghaddam, A novel operational model for interconnected microgrids participation in transactive energy market: a hybrid IGDT/stochastic approach, IEEE Trans. Ind. Inf. 17 (6) (2020) 4025–4035.
[50] F. Moret, T. Baroche, E. Sorin, P. Pinson, Negotiation algorithms for peer-to-peer electricity markets: computational properties, in: 2018 power systems computation conference (PSCC), IEEE, 2018, pp. 1–7.
[51] M. Daneshvar, B. Mohammadi-Ivatloo, M. Abapour, The possibility of using blockchain based cryptocurrency in transactive energy markets: ongoing activities and opportunities ahead, The fourth international energy management and technology conference, 2018.

CHAPTER

5

Modernized P2P energy trading market model and platform for net-zero carbon energy networks

5.1 Introduction

Generally, the present structures of energy networks are aging and have been built based on the centralized energy generation architecture that supports a one-way energy flow [1]. This structure of energy grids relies on producing energy in some specific centers, transmitting it over long distances with considerable losses, and finally distributing it to consumer sites where a high percentage of consumers are passive [2]. In addition to energy losses, transmitting energy over long distances hinders the integration of energy demand and supply in real-time causing overgeneration of energy [3]. The motivations for going toward the decentralized energy system are not just limited to the mentioned reasons and economic and environmental concerns of centralized systems are also enhanced the popularity of decentralized ones. Such issues have accelerated the movement toward the decentralized energy generation paradigm that brings different advantages to the renewable-based system [4,5]. In this regard, the diverse types of distributed energy resources (DERs) have emerged resulting in dramatically changing the energy production, transmission, and delivery that inspire this fact the energy will no longer come from central plants [6]. DERs are the agglomeration of a variety of renewable and non-renewable energy systems along with hybrid energy storage units. However, their non-renewable types have not kept their popularity due to environmental problems, non-permanence of their fuel sources, and economic concerns [7].

Emerging Transactive Energy Technology for Future Modern Energy Networks
DOI: https://doi.org/10.1016/B978-0-323-91133-7.00011-9

Hence, energy networks widely turned on renewable types of DERs that provide environmentally friendly conditions and social-economic benefits. Although renewable energy resources (RERs) are economically efficient and assist the energy structure to become the carbon-zero system, they are stochastic and uncertain in nature with no confident inherently way of managing the balance of energy supply and demand [8]. However, modern energy networks are aimed to be full RERs for producing a clean version of different energy carriers and achieving the net-zero carbon energy networks goal [9]. In this state, how the technology will support modern energy grids in reaching the mentioned goal considering the intermittency of RERs in energy generation? This was the key question that one of its responses was the development of energy trading solutions, in particular, one is the peer-to-peer (P2P) energy trading method. P2P energy trading is one of the energy-sharing mechanisms that allows the market's peer to directly trade energy with another peer without an extra intermediary [10]. However, because the grid conditions are changing under the modernization objectives, the models and platforms that were already for the previous structure of energy grids need to be changed and modernized for the new net-zero carbon structure of energy networks. These changes are essential for the P2P energy trading mechanism and its platforms due to the necessity for the presence of new market players and technologies in the modern energy network. This is the main reason for writing this chapter to analyze the mechanisms and platforms in accordance with the requirements of future modern net-zero carbon energy networks.

5.2 Peer-to-Peer energy trading model

Recently, the rapid increment in the decentralization of the energy generation sector by deploying various types of distributed energy units has dramatically changed the energy distribution procedure [11,12]. In this regard, RERs are found great popularities for doing responsibilities in clean multi-energy generation due to their abilities in providing net-zero carbon and economically efficient conditions for the system. Although these types of distributed energy systems produce 1 percent of global power demand, their adoption rate is increasingly growing, which resulted in affecting the topology of energy grids [13]. This effect appears in diverse opportunities and challenges that both inspire the need for transforming the structure of energy grids to be suitable for new changes. Reducing the dependency on the main grid can be known as the foremost opportunity coming from installing renewable units for energy production [14]. In this respect, a large number of companies and consumers are going towards power self-production. Indeed, conventional consumers are becoming prosumers enabling them to be active players instead of passive ones [15]. As prosumers

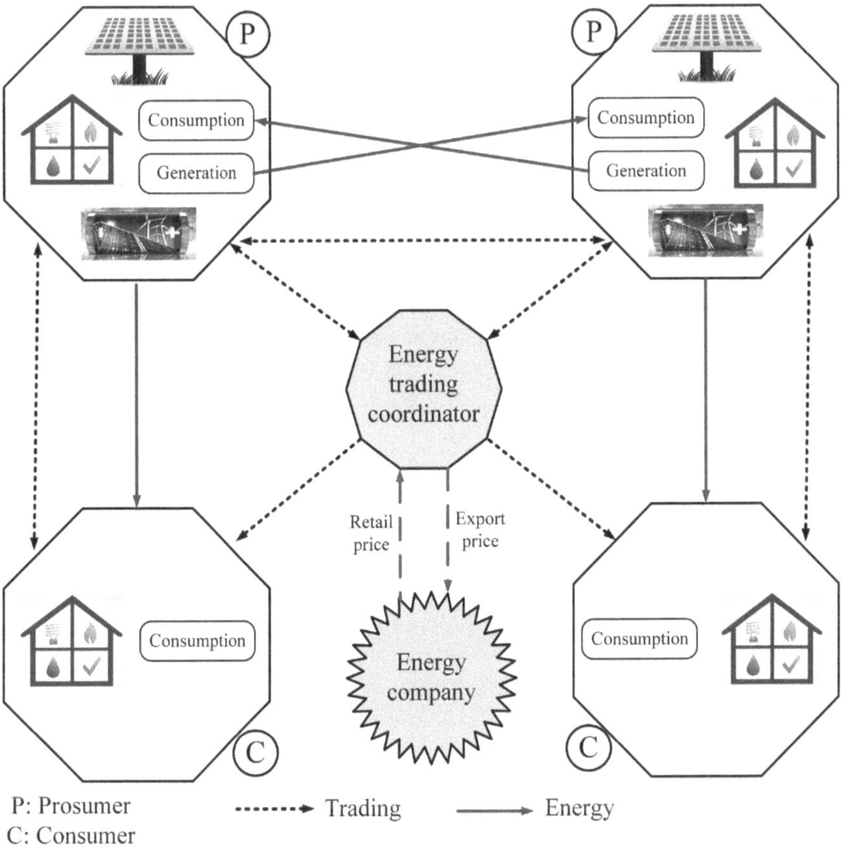

FIGURE 5.1 Schematic of the P2P energy sharing model [19].

are scheduled to install a high degree of RERs for power production due to their higher positive influences in terms of quantity of sunshine and temperature, their power outputs are fluctuating and difficult to predict [16,17]. When prosumers face excess energy in their production, they can analyze different options for energy management. The surplus energy can be exported to the main grid, stored in the storage system, or sold to consumers. Despite the mentioned options are the energy management tools that bring flexibility for prosumers, energy sharing possibility is also offered as another creative way of flexibility for the system. The P2P trade was the name that was given to direct energy sharing between consumers and prosumers [18]. Here, the P2P energy trading model is developed to present a systematic view of P2P energy transactions among multifarious market players. The schematic for the P2P energy sharing model is depicted in Fig. 5.1 [19].

According to Fig. 5.1, four actors are the main parts of the P2P energy sharing model including the energy trading coordinator, energy company, prosumers, and consumers [14]. As this Fig. also clearly indicates the difference between consumers and prosumers, prosumers can generate and consume energy while the consumers only can have consumption action in the system. The trading and energy arrows denote the money and energy exchanges between consumers and prosumers. Here, the energy trading coordinator is responsible for managing entire trading negotiations that are conducted among different peers in the system. The trading arrows present the state of energy transactions, in which the one direct arrow from the coordinator to the consumer states that it can only receive energy from the coordinator. On the other hand, the bidirectional arrows indicate the possibility of buying or selling energy between the prosumers and the energy trading coordinator [19-21].

The P2P energy trading concept is also recognized as a shared economy that the local system is its typical destination for implementation [22]. Indeed, a decentralized model of P2P is also known as the P2P economy, in which the energy interaction among two individuals can be energy sell or buy forms without the interference of any intermediary third-party. In this decentralized model, energy consumers, generators, and prosumers can directly participate in the energy interactions with each other without intermediating third-party entities. In energy interactions, the energy export price typically is intended lower than the retail price with the aim of encouraging prosumers for balancing energy through managing distributed production units [14]. This mechanism drives prosumers to consider their surplus produced energy for sharing with their neighbors to not only reduce dependency on the upstream grid but also gain more benefits while balancing energy in the system. In the P2P energy sharing process, the offers and bids of prosumers and consumers are collected at a domestic price and in the local energy market. Afterward, the collected information is traded with the retailer. In this regard, between the retail price and the export price is the area of setting the domestic price.

The modernized P2P energy trading model allows all participants to share different types of energy in the P2P energy sharing market under the market guidelines. The modernized system is expected to be scalable to welcome any players that would like to be involved in the energy market interactions. It should be adaptable in the high range to be able to cover the emerging technologies and new platforms in line with the system's requirements. Another important thing in the modernized model is its extensibility. Indeed, the market structure needs to be extensible in practice as the population of peers is rapidly increasing day by day along with developing the device technologies that highlight the necessity of being extensible for the modern energy markets. Traditional energy trading systems are mainly structured based on one-way communication. In such a structure, we just have an energy flow that descends from centralized energy production units at the upstream level of the grid to the consumers located at the downstream level of the network. This transaction is usually done over long distances and the cash flow goes another way around. However, what thing expected for energy transactions in the modern structure is multi-directional energy sharing in the local geographical area. The tremendous increment in energy prosumers can affect this structure and procures a more open and decentralized energy grid [23]. Future modern energy grids are planned for the installation of 100 percent RERs to give the most possible clean power for the system. On the other hand, they will be the integrated state of the energy structure to consider multi-energy interactions in the system [24]. Such plans for modern energy grids indicate they will be the host of numerous intelligent devices along with the information, communication, and other related technologies for implementing relevant schemes for achieving the targeted goals. Hence, all models for these energy networks, including energy trading ones, need to be a highly coordinated self-optimization paradigm to effectively handle a set of complex devices and technologies. The P2P energy trading model in the modernized energy structure should be viable in vastness as future modern energy grids will be the incorporation of several energy networks. Indeed, the P2P energy trading model should support the system in providing the required level of reliability and keeping its sustainability when it benefits from the large portion of RERs in energy generation. Thus, it is expected that the P2P energy trading model can facilitate the optimal integration of RERs by opening a new door for flexibility by establishing energy sharing possibilities among peers under modern energy market guidelines. These key attributes are illustrated in Fig. 5.2. Consequently, the P2P energy trading models need to comprise the mentioned attributed in their evolving process making them suitable for the development of future modern multi-energy networks with full or high penetration of RERs that are uncertain in their nature.

5.3 Peer-to-Peer energy trading platform

P2P energy trading platform is the new way of energy management and effective energy sharing strategy that enables proactive customers for energy transactions in the system. The platform supports P2P energy sharing models in realizing multiple key benefits such as eliminating grid losses, energy consumption costs, and peak load demand [25]. The P2P energy sharing platform covers energy selling and buying processes allowing the storage of all information related to contractual relations among participants, energy consumption, and generation. In the traditional process, validation of a digitally transacted energy by a central entity

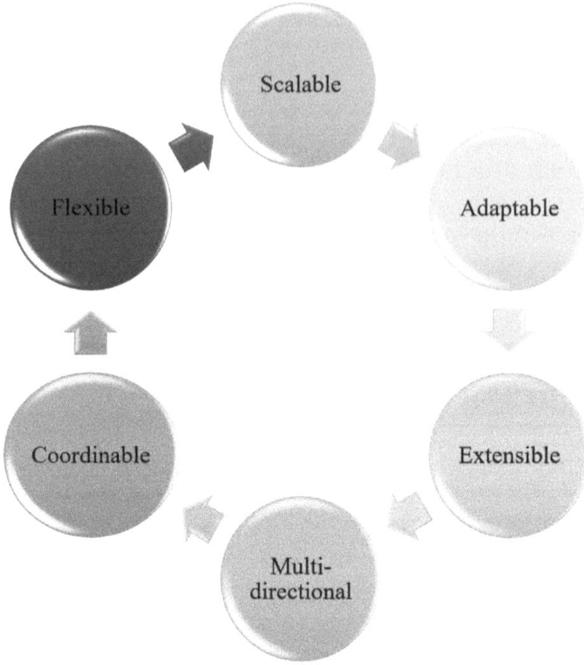

FIGURE 5.2 Some of the key features of modernized P2P energy trading models.

is the inevitable step of completing the related energy sharing process [14]. Hence, the system includes the intermediary for trusting the security of the operation. Currently, energy-sharing platforms not only are in the traditional form, in which transactions need an intermediary for completion, but also another one without requiring the central entity. In the competitive energy environment, secure and reliable platforms are essential for P2P energy-sharing models to provide a confident area of energy exchange for different users. Due to this, studies in the context of the P2P energy sharing platforms are mostly based on testing scalability and security as well as platform architectures. Reference [26] proposed an open-source P2P energy sharing platform to facilitate the energy exchange among various peers by controlling and monitoring self-produced energy at a remote location as well as providing real-time data acquisition. A private Ethereum blockchain is used in the web interface where the energy sharing activities are conducted based on the smart contract and controlling and monitoring of self-generated systems are done by the internet of things platform. Given the key role of the P2P energy sharing mechanism in digitalization, decentralization, and decarbonization of the energy supply chain, the authors in [27] proposed a novel decentralized platform for the P2P energy sharing market using blockchain technology. The proposed platform comprises two blockchain and market layers and is developed to address key challenges that hinder the integration of blockchain and P2P energy sharing mechanisms including the retention of a fair balance between information privacy and economic efficiency, the need for new P2P market designs for accounting inter-temporal dependencies of storage devices, the necessity for the practical implementation of blockchain-based P2P energy sharing system for the fraud-resilient, secure, and efficient energy trading along with removing any other intermediaries' costs. In [28], new models of P2P energy exchanging platforms are investigated for the power sector to clarify the incentives for investing in distributed domestic generation systems that are commonly based on renewable energy sources. A local trading system for electrical energy is modeled for designing it as the dealing platform with the aim of determining the energy selling and purchasing prices on a continuous time basis. The substantial recent interest in the application of blockchain technology and distributed ledger in the local energy sharing platforms for P2P energy sharing has motivated the authors in [29] to procure a new framework for analyzing the distribution network impacts of P2P energy exchanging. To this end, a new methodology is presented for the co-simulation of local P2P energy sharing platforms and electricity distribution networks, in which a distributed double auction trade mechanism empowered by blockchain technology is used by the distribution system simulator that plays an interface role in the P2P energy sharing platform. Attending to the need for privacy, security, and trust in energy sharing among several prosumers has motivated the authors in [30] to propose a secure and efficient P2P energy sharing platform to effectively address the challenges of existing pricing schemes as well as the privacy-preserving behavior of the actual energy consumers in the trading processes. To effectively tackle the aforementioned challenges, a two-layered secure P2P energy trading model is proposed based on the blockchain with two layers including secure energy exchanging as a new consensus mechanism for minimizing the number of malicious validators and authentication for protecting the suggested model from a mutual authentication process and impersonation attacks. In order to alleviate energy imbalances in the production of different distributed users, reference [31] proposed a P2P energy exchanging platform based on blockchain technology as an emerging solution for reliably transacting P2P energy without intending any centralized broker intervention. Due to this, Ethereum's smart contract is used for implementing the P2P energy sharing platform to establish reliable energy trading, a mobile application, and a web page with the aim of monitoring the transaction information such as energy prices and transaction details for enhancing users' accessibility. Reducing the dependability of the energy system on the fossil fuel-based driven main grid is considered as the main objective of proposing the P2P energy trading model in [32]. To identify the objective of different participants in the energy context, a game-theoretic method is proposed for motivating prosumers to conduct P2P energy sharing in both grid-connected and islanded modes, and an energy allocation policy is also used to ensure all prosumers do not exit the energy market without sharing irrespective of the balance among the energy demand and supply. In [33], the authors proposed a new P2P energy sharing model based on the transactive energy concepts to provide appropriate management conditions for the multi-carrier energy in the integrated energy hubs. In this research, as all hubs are equipped with renewable energy systems, uncertainties are modeled by applying the stochastic programming method and the P2P energy sharing opportunity is used for adding energy trading as the suitable tool for balancing multi-energy of hubs in the optimal scheduling problem. The Ethereum blockchain technology is used in [34] for presenting the P2P energy trading platform along with the dedicating components and architecture aiming to realize the trustful, secure, and transparent energy sharing processes. In this system, stakeholders provide the customers of a P2P energy trading model that can exchange clean energy and consume it. To provide a high level of controllability and efficiency by participating aggregated DERs in the ancillary services, the microgrid is organized into several nanogrids in [35] to effectively utilize energy trading possibilities in the system. For this aim, a multi-market nanogrid exchanging paradigm presented in the mentioned reference to simultaneously eliminate the frequency regulation procurement and real-time imbalance based on the communication and P2P energy sharing architecture as well as participating in the three consecutive markets including ancillary service market, balancing market, and P2P bilateral energy market. With rapid development in urbanization, decentralization, digitalization, and decarbonization are recognized as three key pillars for supplying the substantial increase in the energy demand that uses energy trading markets for the global low-carbon economy. Given this issue, the authors in [36] analyzed the positive roles of electric vehicles' interactive transaction behaviors as well as buildings in creating a sustainable transactive energy community considering human social space, data cyber space, and energy physical space. A hierarchical P2P-energy transaction model is presented in [37] that divides energy characteristics into green and normal energy considering the preferences of renewable energy. In this research, the energy transaction problem is solved by presenting a hierarchical method, in which prosumers' goals and energy transactions are intended based on the assumption that both consumers and prosumers can exchange energy with each other along with receiving green and normal energy from the power grid. Moreover, a P2P energy trading system is proposed in [38] to establish P2P energy sharing among dwellings for their optimal energy management when they are equipped with micro fuel cell combined power and heat units. In this research, a distributed alternating direction method of multipliers (ADMM) technique is applied to solve the convexified problem that was already non-convex due to the nonlinearity comes from the dependency of heat and power efficiencies on the output power of the fuel cell units.

To sum up, recent literature indicates four major methods as the most prominent contributions in advancing the present P2P energy sharing platforms towards the modernized energy exchanging platforms. Herein, blockchain technology, constrained optimizations, game theory, and auction-based theory are four typical techniques that are exerted in different problems in recent literature [39]. The goal of blockchain technology is to establish a required data format in P2P energy grids for users that disseminate and replicate data with the aim of facilitating decentralize, transparent, and secure energy exchange. A mathematically programmed technique is implemented by the constrained-based optimization for optimizing parameters of P2P energy exchanging under the different types of market restrictions. The best mathematical way is labeled the game theory for investigating the decisions of particular individuals in a competitive and plausible setting, in which the actions of one participant can affect the actions of others. Enabling a group of P2P market purchasers and distributors for energy sharing in a sequential manner is the objective of the auction-based theory that does this by keeping track of their collaborations in the marketplace.

5.4 Summary

Growing criticisms of the operation of conventional energy systems empowered by fossil fuels have caused the flourishing of the idea of developing net-zero carbon energy networks. Although this type of energy grid offers net-zero energy production processes without harmful pollutants, they have a long distance for realization in the practice due to the complete resilience to renewable energy systems that have stochastic outputs. For such grids, energy trading methods are introduced as one of the strong solutions to meet a significant portion of the grid's flexibility requirements. In this regard, a P2P energy sharing method has emerged with various models and platforms to enable diverse participants of the energy market to energy exchange for different goals. However, the remarkable technological advancements in line with the grid modernization have opened the area of presence of new innovative technologies that creates the need for a modernized model and platform of the P2P energy sharing for future modern energy grids. Due to this, this chapter scrutinized the modernized P2P energy trading model and platform for future modern net-zero carbon energy networks. The P2P energy trading model was investigated by providing the related structure for the P2P energy trading considering modernized factors in accordance with the grid modernization objectives. Afterward, the key features of the modernized P2P energy sharing model were discussed to give a clear overview of future P2P energy trading models for modern multi-energy grids. Finally, the P2P energy trading platform was analyzed by evaluating the recently developed platforms and their attributes to allow for the P2P platform advancement under the modernized guidelines. Indeed, the modernized P2P energy sharing model and platform can enable future modern net-zero carbon multi-energy networks to be enough flexible in front of the full level of renewable energy units.

5.5 Questions

The careful analyzing whole of this chapter is expected to enable readers to effectively evaluate the following questions.

1) Why the modernized P2P energy sharing model and platform are necessary for future modern net-zero carbon energy networks?
2) How the P2P energy trading model can be developed in line with the grid modernization programs for future modern energy grids?
3) How the P2P energy sharing model can facilitate energy tradings among different participants of the energy market?
4) What are the key features of modernized P2P energy trading models and why they are key in the modernization process of P2P energy exchanging models?
5) What are the typical platforms for the P2P energy trading method and each of them is famous for which properties?
6) How the modernized P2P energy trading platform can support the system for efficient energy management in the renewable-based grid area?

5.6 Suggestions

Given the topic of this chapter, the following suggestions are provided for the readers assisting them to pursue related research clues.

1) Modernizing the P2P energy sharing model and platform can ease the condition for the presence of this energy trading method in the body of energy management schemes of future modern net-zero carbon energy networks.
2) Designing the P2P energy trading model in accordance with the grid modernization objectives can smooth the pass for flexibility improvement mostly based on the energy sharing possibilities.
3) Developing the P2P energy sharing platform can increase the ability of future modern net-zero carbon energy networks to optimally control and manage multi-energy interactions, especially in the presence of 100 percent renewable energy units.

5.7 Future trends

Future trends can give a proper overview for the readers to form their related works and develop effective solutions for remaining gaps, which are listed as follows for this chapter.

1) Modernizing the P2P energy sharing model and platform is the necessary step for the contribution of the P2P energy trading method in structuring the direct energy exchanges among various participants of the modern energy market in future net-zero carbon energy grids.
2) The modernized P2P energy trading model is a crucial part of investing in the energy sharing techniques for meeting the sufficient flexibility of future modern energy grids.
3) The P2P energy trading platform needs to be modernized to be an appropriate option for incorporating into the energy management and control schemes of modern multi-energy networks with 100 percent renewable systems.

References

[1] M.T. Hagh, H. Rezaei, M. Daneshvar, Faulted feeder identification in active grounded networks, IET Gener. Transm. Distrib. 13 (15) (2019) 3476–3483.

[2] M. Daneshvar, M. Abapour, B. Mohammadi-ivatloo, S. Asadi, Impact of optimal DG placement and sizing on power reliability and voltage profile of radial distribution networks, Majlesi J. Electr. Eng. 13 (2) (2019) 91–102.

[3] G. Vieira, J. Zhang, Peer-to-peer energy trading in a microgrid leveraged by smart contracts, Renew. Sustain. Energy Rev. 143 (2021) 110900.

[4] W.H. Liu, et al., Development and optimization of an integrated energy network with centralized and decentralized energy systems using mathematical modelling approach, Energy 183 (2019) 617–629.

[5] M. Daneshvar, E. Babaei, Exchange market algorithm for multiple DG placement and sizing in a radial distribution system, J. Energy Manage. Technol. 2 (1) (2018) 54–65.

[6] A.S. Farsangi, S. Hadayeghparast, M. Mehdinejad, H. Shayanfar, A novel stochastic energy management of a microgrid with various types of distributed energy resources in presence of demand response programs, Energy 160 (2018) 257–274.

[7] J. Faraji, H. Hashemi-Dezaki, A. Ketabi, Multi-year load growth-based optimal planning of grid-connected microgrid considering long-term load demand forecasting: a case study of Tehran, Iran, Sustain. Energy Technol. Assess. 42 (2020) 100827.

[8] M. Daneshvar, B. Mohammadi-ivatloo, S. Asadi, S. Galvani, Short term optimal hydro-thermal scheduling of the transmission system equipped with pumped storage in the competitive environment, Majlesi J. Electr. Eng. 14 (1) (2020) 77–84.

[9] M., Daneshvar, S., Asadi, and B., Mohammadi-Ivatloo, "Grid Modernization– Future Energy Network Infrastructure," 2021.

[10] G. Li, Q. Li, X. Yang, R. Ding, General Nash bargaining based direct P2P energy trading among prosumers under multiple uncertainties, Int. J. Electr. Power Energy Syst. 143 (2022) 108403.

[11] A.T. Pham, L. Lovdal, T. Zhang, M.T. Craig, A techno-economic analysis of distributed energy resources versus wholesale electricity purchases for fueling decarbonized heavy duty vehicles, Appl. Energy 322 (2022) 119460.

[12] M. Daneshvar, B. Mohammadi-Ivatloo, K. Zare, A. Anvari-Moghaddam, Transactive energy strategy for energy trading of proactive distribution company with renewable systems: a robust/stochastic hybrid technique, e-Prime-Advances in Electrical Engineering, Electron. Energy 2 (2022) 100028.
[13] C.C. IRENA, R. Energy, National policies and the role of communities, cities and regions (Report to the G20 Climate Sustainability Working Group (CSWG)), International Renewable Energy Agency, Abu Dhabi (2019).
[14] E.A. Soto, L.B. Bosman, E. Wollega, WD. Leon-Salas, Peer-to-peer energy trading: a review of the literature, Appl. Energy 283 (2021) 116268.
[15] Y. Luo, S. Itaya, S. Nakamura, P. Davis, Autonomous cooperative energy trading between prosumers for microgrid systems, in: 39th annual IEEE conference on local computer networks workshops, IEEE, 2014, pp. 693–696.
[16] M. Daneshvar, H. Eskandari, A.B. Sirous, R. Esmaeilzadeh, A novel techno-economic risk-averse strategy for optimal scheduling of renewable-based industrial microgrid, Sustain. Cities Soc. 70 (2021) 102879.
[17] S. Ding, Z. Tao, R. Li, X. Qin, A novel seasonal adaptive grey model with the data-restacking technique for monthly renewable energy consumption forecasting, Expert Syst. Appl. 208 (2022) 118115.
[18] K.Y. Bandara, S. Thakur, J. Breslin, Flocking-based decentralised double auction for P2P energy trading within neighbourhoods, Int. J. Electr. Power Energy Syst. 129 (2021) 106766.
[19] Y. Zhou, J. Wu, C. Long, Evaluation of peer-to-peer energy sharing mechanisms based on a multiagent simulation framework, Appl. Energy 222 (2018) 993–1022.
[20] A. Roy, A. Bruce, I. MacGill, The potential value of peer-to-peer energy trading in the Australian national electricity market, in: Asia-pacific solar research conference, 2016, 2016.
[21] M.U. Gurmani, et al., Energy trading between prosumer and consumer in p2p network using blockchain, in: International Conference on P2P, Parallel, Grid, Cloud and Internet Computing, Springer, 2019, pp. 875–886.
[22] J. Hamari, M. Sjöklint, A. Ukkonen, The sharing economy: why people participate in collaborative consumption, J. Assoc. Inf. Sci. Technol. 67 (9) (2016) 2047–2059.
[23] Z. Li, S. Bahramirad, A. Paaso, M. Yan, M. Shahidehpour, Blockchain for decentralized transactive energy management system in networked microgrids, Electr. J. 32 (4) (2019) 58–72.
[24] Y. Li, B. Wang, Z. Yang, J. Li, C. Chen, Hierarchical stochastic scheduling of multi-community integrated energy systems in uncertain environments via stackelberg game, Appl. Energy 308 (2022) 118392.
[25] S. Rezaei, A. Ghasemi, Stochastic scheduling of resilient interconnected energy hubs considering peer-to-peer energy trading and energy storages, J. Energy Storage 50 (2022) 104665.
[26] M.J.A. Baig, M.T. Iqbal, M. Jamil, J. Khan, Design and implementation of an open-Source IoT and blockchain-based peer-to-peer energy trading platform using ESP32-S2, Node-Red and, MQTT protocol, Energy Rep. 7 (2021) 5733–5746.
[27] A. Esmat, M. de Vos, Y. Ghiassi-Farrokhfal, P. Palensky, D. Epema, A novel decentralized platform for peer-to-peer energy trading market with blockchain technology, Appl. Energy 282 (2021) 116123.
[28] T. Cortade, J.-C. Poudou, Peer-to-peer energy platforms: incentives for prosuming, Energy Econ. 109 (2022) 105924.
[29] B.P. Hayes, S. Thakur, J.G. Breslin, Co-simulation of electricity distribution networks and peer to peer energy trading platforms, Int. J. Electr. Power Energy Syst. 115 (2020) 105419.

[30] A.S. Yahaya, N. Javaid, A. Almogren, A. Ahmed, SM. Gulfam, A. Radwan, A two-stage privacy preservation and secure peer-to-peer energy trading model using blockchain and cloud-based aggregator, IEEE Access 9 (2021) 143121–143137.

[31] S. Kwak, J. Lee, Implementation of blockchain based P2P energy trading platform, in: 2021 International Conference on Information Networking (ICOIN), IEEE, 2021, pp. 5–7.

[32] W. Amin, et al., A motivational game-theoretic approach for peer-to-peer energy trading in islanded and grid-connected microgrid, Int. J. Electr. Power Energy Syst. 123 (2020) 106307.

[33] M. Daneshvar, B. Mohammadi-Ivatloo, K. Zare, A novel transactive energy trading model for modernizing energy hubs in the coupled heat and electricity network, J. Cleaner Prod. 344 (2022) 131024.

[34] D. Rangelov, B.S.K. Subudhi, P. Lämmel, M. Boerger, N. Tcholtchev, J. Khan, Design and Specification of a Blockchain-based P2P Energy Trading Platform, in: 2021 IEEE 21st International Conference on Software Quality, Reliability and Security Companion (QRS-C), IEEE, 2021, pp. 636–643.

[35] Q. Hu, Z. Zhu, S. Bu, KW. Chan, F. Li, A multi-market nanogrid P2P energy and ancillary service trading paradigm: mechanisms and implementations, Appl. Energy 293 (2021) 116938.

[36] Y. Wu, Y. Wu, J.M. Guerrero, J.C. Vasquez, Decentralized transactive energy community in edge grid with positive buildings and interactive electric vehicles, Int. J. Electr. Power Energy Syst. 135 (2022) 107510.

[37] D.-H. Park, Y.-G. Park, J.-H. Roh, K.Y. Lee, J.-B. Park, A hierarchical peer-to-peer energy transaction model considering prosumer's renewable energy preference, IFAC-PapersOnLine 52 (4) (2019) 312–317.

[38] D.H. Nguyen, T. Ishihara, Distributed peer-to-peer energy trading for residential fuel cell combined heat and power systems, Int. J. Electr. Power Energy Syst. 125 (2021) 106533.

[39] H. Javed, H. Muqeet, M. Irfan, Recent Trends, Challenges and Future Aspects of P2P Energy Trading platforms in Electricalbased Networks Considering Blockchain Technology: a Roadmap Towards Environmental Sustainability. Front, Energy Res. 10 (2022) 134.

CHAPTER

6

New emerging transactive energy technology in modernizing future hybrid energy networks

6.1 Introduction

Over the last years, energy trading has become a most popular strategy especially when renewable energy sources (RESs) are intended to be a key part of the energy generation sector [1]. The popularity of RESs is more identified after the primary efforts have been done to use distributed generators in the distribution system to promote decentralized energy generation [2]. This issue mostly backs to the addition of RESs led by the combination of technological advances and policy directives [3]. The power system faces huge uncertainties when it relies on RESs to give clean power output. However, creating massive flexible capacities is necessary to deal with stochastic changes in RESs outputs [4]. Indeed, as uncertainties are expected in the energy generation of RESs, the grid requires adequate flexibility to reliably act in the energy management of the system. In this regard, energy trading methods can procure sufficient flexibility for the system in the presence of RESs [5]. This key role of the energy trading mechanism in facilitating the adoption of RESs has resulted in appearing multifarious energy exchanging strategies with different properties. Transactive energy (TE) is known as one of these strategies that is developed to be the main tool for modernizing future energy grids [6]. A set of mechanisms are targeted especially in terms of economics and control to pose a procedure for creating a dynamic power balance between the power supply and demand. These mechanisms are defined mostly based

Emerging Transactive Energy Technology for Future Modern Energy Networks
DOI: https://doi.org/10.1016/B978-0-323-91133-7.00002-8

on various factors as the value such as emission or money. In this regard, TE systems are appeared to support the realization of the TE goals [7]. They are for designing and operating the innovative energy-sharing architecture that makes the electrification process easy. Accessing such an electrification architecture is crucial that gives energy exchanging opportunities for the system to help it to keep its sustainability, reliability, and flexibility for secure energy supply. As all the mentioned achievements are necessary for future modern energy networks (FMENs) that are targeted for 100 percent carbon-free energy generation, TE systems can play the main supportive role in earning such advantages. This can be mostly carried out by establishing a distributed information system that supports the grid to encompass the interoperability among the energy market's participants. Therefore, an effective remedy is essential to benefit the TE for addressing the challenges in the modern electrification of FMENs with a full/high share of RESs. This chapter is concentrated on clarifying the TEs role in modernizing future energy grids. The definition of TE is provided to effectively highlight the basic goals for the TE. What are features of the TEs framework is another key question intended in this chapter. This work also covers challenges, opportunities, principles, and strategies of the TE in energy interactions of FMENs. Eventually, the chapter investigates the capability of TE technology in modernizing future multi-carrier energy networks to give an appropriate overview of energy sharing models in the modern structure of the system.

6.2 TE definition

In recent years, the significant proliferation of RESs at bulk power and decentralized energy generation units at grid-edge and distribution levels give rise to operational issues of the power system in both distribution and transmission sectors [8]. At the bulk transmission level, such evolutions create the requirement for higher levels of ancillary services as well as various types of network services such as synthetic damping and inertia, flexibility reserves, etc. This is while the distribution operational level is expected to need protection schemes and conventional feeder relief as well as voltage support, grid-edge reactive, and phase balancing as the new grid services. With increasingly emerging intelligent devices and controllable components in the body of the power grid, the call for advanced energy management and control mechanisms has become a critical step for achieving capable models of monitoring and adjusting the system with diverse loads considering the external or local signals. This requirement is intensified since utilizing a vast range of RESs has become one of the overarching goals of FMENs. To address the similar challenges ahead of modernizing U.S. electric power system, the U.S. Department of Energy (DOE)

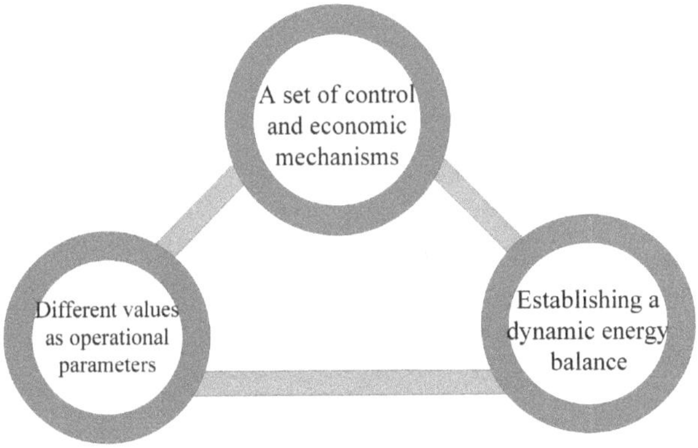

FIGURE 6.1 Three key aspects of the TE technology.

funded the Pacific Northwest National Laboratory (PNNL) to focus on one of the first practical solutions of what later was called TE [9].

The TE is defined by the GridWise Architecture Council (GWAC) as "a set of economic and control mechanisms that allows the dynamic balance of supply and demand across the entire electrical infrastructure using value as a key operational parameter" [10]. This definition indicates that TE typically refers to three key aspects that are illustrated in Fig. 6.1.

Based on the mentioned definition, TE relies on the set of control and economic mechanisms aiming to support energy transitions toward affordable and sustainable structures. The TE offers local energy exchanging environments for bringing enough flexibility that requires rigorously re-engineering of the energy network infrastructure [11]. Such TE plans cover another aspect of TE in creating a dynamic energy balance, especially in the presence of intermittent sources. This goal can also be realized by adopting various strategies of the TE that each of them can open a special door for energy exchanging of different market participants [12]. On the other hand, such processes are designed to form suitable conditions for the system to pursue different goals that are taken into account as the values in the TE area. Indeed, according to the TE definition, TE considers various values as key operational parameters. These values can be minimizing emissions and maximizing profits or minimizing costs. However, with emerging new technologies in different sciences alongside substantial changes in the network topology, the definition of TE provided by the GWAC cannot provide a clear scope and comprehensive overview for TE schemes in line with the grid advancements. These shortcomings are briefly demonstrated in Fig. 6.2.

Given Fig. 6.2, the initial definition of TE only covers the economic and control mechanisms that limit the emergence of other innovative mechanisms with the most applicable features. On the other hand, limited

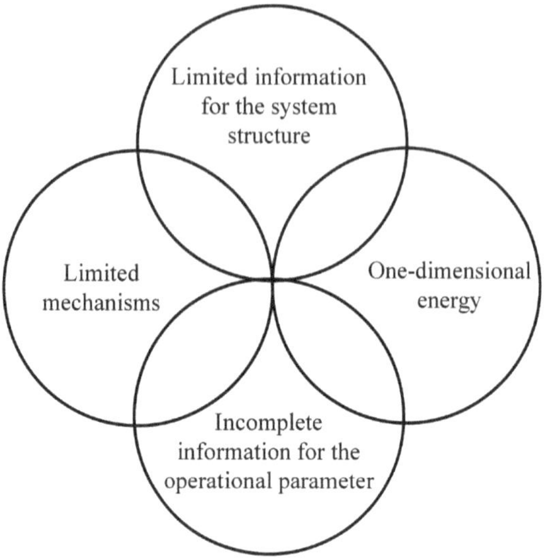

FIGURE 6.2 Shortcomings of the TE definition.

information is provided for the system structure in creating a dynamic energy balance and how it can be established in different structures. The definition is focused on one-dimensional energy while plenty of multi-carrier energy units is appeared in the grid due to recent developments in hybrid energy systems [13]. Moreover, limited information is given regarding the key operational parameters that make the related process remain vague. All of the aforementioned reasons pose the need for a complete definition that can effectively identify the scope of TE in FMENs. The new definition should include information to welcome the innovative solutions for effectively tackling the prominent challenges of FMENs. To this end, the authors in [14] proposed a new definition for the multi-carrier transactive energy considering the different parts of the mentioned gaps. The new definition describes the TE as "a set of economic, social, environmental, and control mechanisms that is responsible to establish a dynamic balance for multi-dimensional energy across the entire multi-carrier energy infrastructure with a full/high share of clean energy resources using economical, technical, social, and environmental targets as the key operational parameters for making the hybrid energy networks more reliable, resilient, flexible, efficient, sustainable, and stable than ever before" [14].

The new definition covers different key mechanisms to facilitate the implementation of systematic plans aiming to enhance social welfare across the modern energy area. It promotes the exploitation of hybrid energy systems by outstretching multi-vector energy units and multi-energy

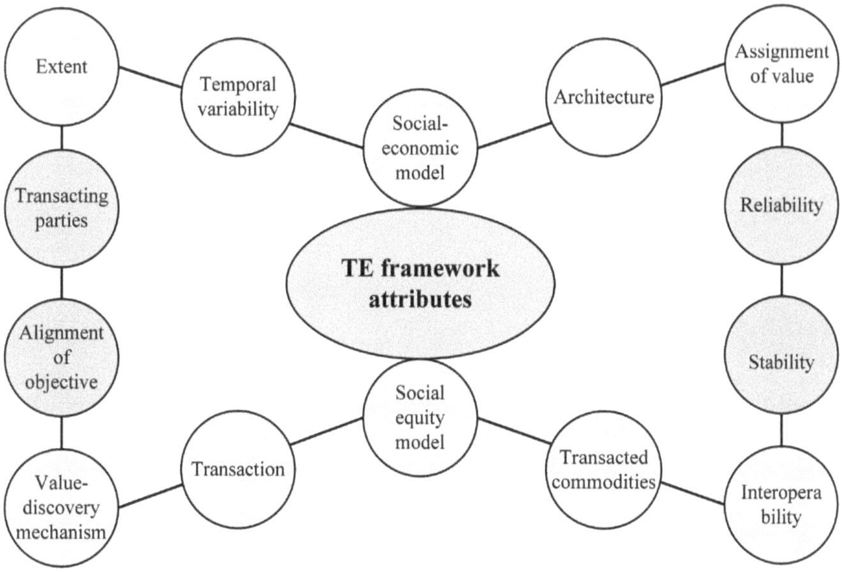

FIGURE 6.3 TE framework attributes.

strategies for balancing energy. It also clarifies the main goals of future modern energy networks that can drive schedulings toward grid modernization.

6.3 TE framework

The TE framework is proposed to procure clear conditions for interoperability among different participants to have an effective consensus in reaching the vital objectives of the integrated system [15]. The TE framework renders a neutral area of energy exchange for all market members such as shareholders, producers, and consumers to make the appropriate negotiating opportunity possible with the aim of addressing the challenges and concerns of the power grid [16]. The overarching goal of designing such a framework is to design significant models for power-sharing among various communities in the TE landscape [17]. Indeed, it promotes the community of practice for all participants of the energy market to drive the grid's goals towards realization. The TE framework carries some specific attributes that are demonstrated in Fig. 6.3.

The TE framework supports social-economic and equity models for designing an innovative architecture being adaptive to the goals of modern energy networks [18]. It supports state-of-the-art technologies for coordinating different parts of the intelligent system by developing a common

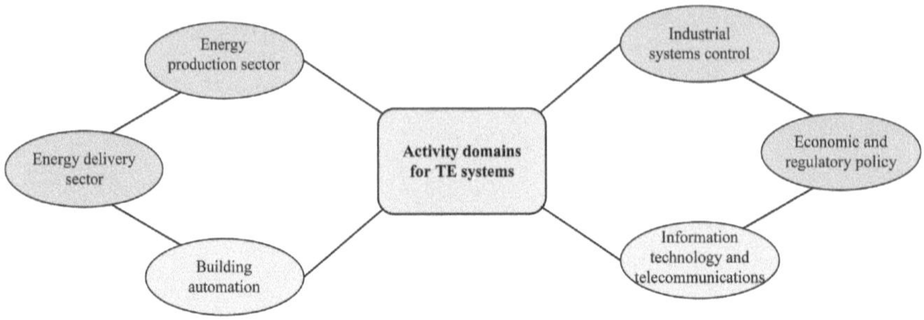

FIGURE 6.4 Activity domains for TE systems.

vocabulary among various parts. On the other hand, the TE framework is for providing efficient energy as well as ensuring the reliability, stability, flexibility, sustainability, and safety of the system under the multifarious energy interactions.

6.4 TE principles

In recent years, a general belief is formed that tomorrow's energy grid will be a collection of sophisticated information and communication technologies for delivering affordable and unbroken clean energy [19]. In this respect, the need for reliable integrating RESs to meet the 100 percent clean energy production goal of FMENs has also increased the complexity of the system in handling energy transactions. Herein, such a complex system with extensive smart components and stochastic resources requires a holistic paradigm to facilitate the implementation of grid modernization schemes. TE is targeted to support this process by empowering the community of practice for interoperability. As the successfulness of such procedures needs to expand activity domains [20], the DOE has enlarged the GWAC focuses to different layers of the modern grid [16]. Fig. 6.4 portrays these activity domains for TE systems [16].

To facilitate the energy activities under TE systems, some key principles are defined by the GWAC during different workshops. The main objective of these principles is to guide the market members to develop their strategies in line with the TE objectives. They are also necessary for developing appropriate models of the TE in the modern energy community. These principles are as follows [9,16].

- Providing a highly coordinated form of self-optimization systems;
- Developing holistic models for maintaining the systems' stability, control, and reliability;
- Designing architectures that are auditable and visible at interfaces;

- Building a fair energy trading environment for market members who are eligible to participate in the market interactions;
- Being adaptable, scalable, and extensible are the mandatory features of TE schemes;
- Following the required standards of the system is one of the key responsibilities of participants in energy transactions.

6.5 TE strategies

The popularity of RESs for clean energy generation has moved the efforts to develop network paradigms for adopting a high level of them that is become as one of the major recent evolutions of energy grids. How developing such frameworks for reliable energy supply is a general concern of the energy area with RESs [21]. The criticality of this challenge relates to the uncontrollability property of RESs in power production that makes the optimal operation of the system difficult. To address this challenge, ambitious endeavors are formed to support the maximum utilization of RESs in the body of the energy generation part. However, achieving targeted and intended results have highlighted the necessity of capable strategies for the resilient operation of the grid. In this regard, the TE strategy is emerged to offer a multitier hierarchical energy market paradigm for allowing unfettered participation of RESs in the decarbonization of FMENs. Indeed, the TE strategy is specialized to play a critical role in completing the interactions of the energy supply chain [22]. Specifically, its structure is designed in a way to facilitate the transformation of the traditional grid into a democratized, digitized, and decentralized network. The TE strategy offers many potential advantages, both to society and individuals. It combines energy and information to enable energetic transactions that are highly coordinated and purposeful interactions [23]. In this respect, several key benefits of the TE strategy are illustrated in Fig. 6.5.

According to Fig. 6.5, the TE strategy is one of the necessary requirements for modernizing FMENs penetrated with RESs. It provides a complete response potential for the flexible systems aiming to coordinately control energy transactions. TE strategy takes appropriate incentive mechanisms into account for evolving an active, efficient, and reliable energy market. All of these capabilities are essential for developing a democratized energy environment to ease the energy interactions of the market's participants.

6.6 TE role in grid modernization

Over recent years, energy networks have witnessed an upward trend of the active participation of decentralized energy generation units as well as

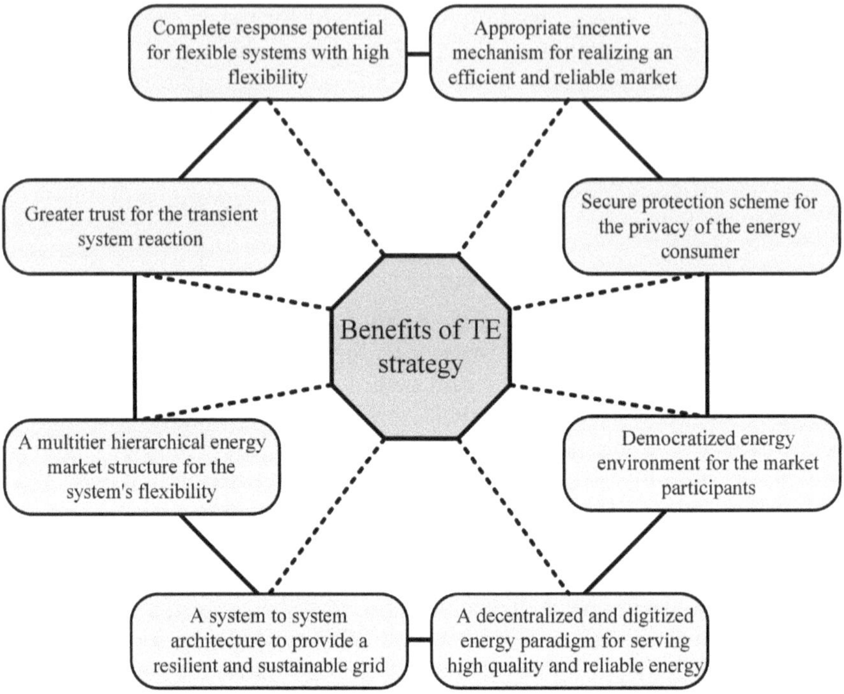

FIGURE 6.5 Benefits of TE strategy.

grid-edge assets in the network [24]. Such transformations have exposed the energy landscape to experience accelerating developments. RESs are supposed to play a central role in this evolution pass especially due to their impress in eco-friendly and affordable energy production. The intermittency feature of RESs imposes huge uncertainties on the system that threaten the sustainable operation of the system. For as much as FMENs are planned to furnish 100 percent RESs to generate carbon-free power, RESs will be an inevitable part of the energy production process. Therefore, deeply scrutinizing their role in the energy network's interactions is a necessary step in grid modernization. In this regard, multifarious systematic strategies are undeniable to overcome the similar challenges of FMENs. To this end, increasing interest is formed in TE systems to offer energy trading solutions for adopting RESs in the large domain by supporting both peer-to-peer and peer-to-market energy sharing [24].

Peer-to-peer energy trading involves bilateral transactions of energy and energy derivatives among market producers and consumers. This is while peer-to-market energy sharing refers to the market-based procurement of grid services by operators or owners of grid-edge assets or their agents. In this regard, TE agents can establish a coordinated hierarchical

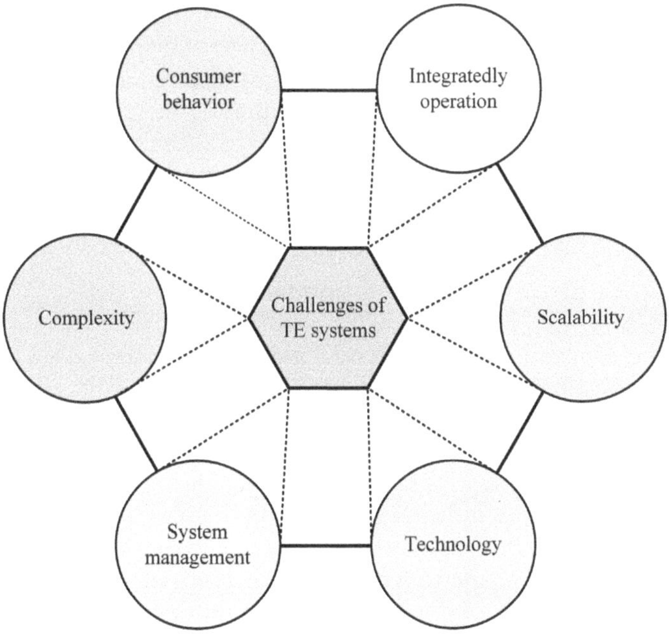

FIGURE 6.6 Challenges of TE systems.

energy interaction processing among the network's components that procures a highly flexible and efficient energy supply in a sustainable manner. This TE architecture is not limited only to the electrical energy grid but also supports multi-carrier energy grids for multi-energy interactions in FMENs. As TE systems and agents enable the system to be developed in line with grid modernization goals, some specific challenges have appeared that slow down the progress of the process. Fig. 6.6 shows these challenges for the TE systems.

According to this Fig., consumer behaviors are one of the challenges that affect the applicability of TE schemes. The different thoughts and opinions on modernizing energy grids as well as various patterns and preferences for consumers' energy consumption are intensified this challenge more than ever before. Another issue backs to the integratedly operation of several energy networks in the TE system. FMENs are aimed to be the coupled structure of several energy grids such as power, gas, heat and cool, and water networks due to the need for considering inevitable dependencies among them for realistic modeling of the system. Simultaneously optimizing several energy networks with numerous constraints under the TE architecture also upsurges the complexity of the system modeling. On the other hand, technological development renders another challenge for coordinately managing a large number of smart devices in the overall system. The scalability of such advancements makes the new necessities for energy management of FMENs that restrict the applicability of some technologies. The aforementioned challenges inspire the need for new TE models to effectively address them for facilitating the grid modernization process. The next chapters of this book try to more highlight the application of TE technology and propose new TE models for addressing the challenges ahead of modernizing future multi-energy grids.

6.7 Summary

Ongoing events in energy networks indicate the urgent need for properly utilizing emerging technologies to effectively address the energy crisis. This crisis is more sensed by enhancing the requests for multi-energy as well as preferring RESs as the main sources of energy for FMENs. How the flexibility of the system is procured for reliable and sustainable energy meeting has become an arduous challenge in the presence of plenty RESs. In this regard, TE technology has appeared as a vigorous solution to overcome the mentioned challenges of FMENs. Indeed, it is recognized as one of the main contenders for implementing grid modernization plans. Given the key role of TE in modernizing future multi-energy grids, this chapter aims at evaluating the TE features and capability for FMENs in detail. The definition of TE is intended in both general and comprehensive versions to give complete information regarding the TE goals and missions. The TE framework is described alongside clarifying the relevant attributes for clearly highlighting TE application for FMENs. In order to justify the usage of TE as the main part of grid modernization, its strategies are elaborated along with concentrating on its benefits for the system. Finally, the vital role of TE systems in modernizing future energy grids is explained and the expected challenges are also discussed to make the future trends of grid modernization clear.

6.8 Questions

The careful analyzing whole of this chapter is expected to enable readers to effectively evaluate the following questions.

1) What is the TE and why it is developed by the GWAC?
2) What are the general and comprehensive definitions of TE?
3) How the TE mission is in line with modern energy grids' goals?
4) What is the key role of energy trading mechanisms in facilitating the adoption of RESs?
5) Why the new definition is suggested for the TE and what are its features?

6) Why TE framework is developed for modernizing future multi-vector energy grids?
7) What are the important attributes of the TE framework?
8) How TE framework can facilitate grid modernization processes?
9) What are the TE principles in managing energy tradings of the system?
10) How TE mechanisms can establish interoperability among market members?
11) What are the activity domains for TE systems with numerous intelligent devices?
12) What are the TE strategies for improving the energy interactions of different participants?
13) What are the advantages of the TE strategies for FMENs?
14) What is the prominent role of the TE in grid modernization?
15) What are the expected challenges ahead of the applicability of TE systems in modernizing future multi-carrier energy networks?

6.9 Suggestions

Given the topic of this chapter, the following suggestions are provided for the readers assisting them to pursue related research clues.

1) Developing TE concepts in accordance with the grid modernization goals can improve its application for modernizing future energy networks.
2) Designing a holistic TE framework can increase its capability in establishing purposeful interoperability among energy trading market participants.
3) Providing complete TE principles can support the innovative models to carry the necessary properties required for transiting from conventional grids to modern energy networks.
4) Developing practical TE strategies can improve energy interactions among different market members aiming to create an energy balance in the system with the penetration of RESs.

6.10 Future trends

Future trends can give a proper overview for the readers to form their related works and develop effective solutions for remaining gaps, which are listed as follows for this chapter.

1) The current TE concepts need to be upgraded in line with the grid modernization objectives aiming to provide appropriate conditions for modernizing future energy networks.

2) An innovative TE framework is necessary to be designed in a way to establish purposeful interoperability among energy trading market participants.
3) A complete version of TE principles needs to be procured to support TE models for successfully realizing FMENs.
4) Effective TE strategies are essential for improving energy interactions as well as developing a community of practice to make the energy structure sustainable and reliable in energy meeting.

References

[1] K. Zhang, B. Zhou, C.Y. Chung, S. Bu, Q. Wang, N. Voropai, A Coordinated Multi-Energy Trading Framework for Strategic Hydrogen Provider in Electricity and Hydrogen Markets, IEEE Trans. Smart Grid (2022).
[2] M. Daneshvar, E. Babaei, Exchange market algorithm for multiple DG placement and sizing in a radial distribution system, J. Energy Manage. Technol. 2 (1) (2018) 54–65.
[3] A. Papalexopoulos, The Evolution of the Multitier Hierarchical Energy Market Structure: the emergence of the transactive energy model, IEEE Electrif. Mag. 9 (3) (2021) 37–45.
[4] M. Daneshvar, B. Mohammadi-Ivatloo, K. Zare, M. Abapour, S. Asadi, A. Anvari-Moghaddam, Chance-constrained scheduling of hybrid microgrids under transactive energy control, Int. J. Energy Res. 45 (7) (2021) 10173–10190.
[5] M. Khorasany, A.S. Gazafroudi, R. Razzaghi, T. Morstyn, M. Shafie-khah, A framework for participation of prosumers in peer-to-peer energy trading and flexibility markets, Appl. Energy 314 (2022) 118907.
[6] M. Daneshvar, B. Mohammadi-Ivatloo, K. Zare, A novel transactive energy trading model for modernizing energy hubs in the coupled heat and electricity network, J. Cleaner Prod. 344 (2022) 131024.
[7] S. Bhattacharya, T. Ramachandran, A. Somani, D.J. Hammerstrom, Impacts of Energy Flexibility in Transactive Energy Systems With Large-Scale Renewable Generation, IEEE Access 10 (2022) 14870–14879.
[8] F. Rahimi, A. Ipakchi, Using a transactive energy framework: providing grid services from smart buildings, IEEE Electrif. Mag. 4 (4) (2016) 23–29.
[9] D.J. Hammerstrom, S.E. Widergren, C. Irwin, Evaluating transactive systems: historical and current US DOE research and development activities, IEEE Electrif. Mag. 4 (4) (2016) 30–36.
[10] R. Ambrosio, Transactive energy systems, IEEE Electrif. Mag. 4 (4) (2016) 4–7.
[11] M. Daneshvar, B. Mohammadi-Ivatloo, K. Zare, A. Anvari-Moghaddam, Transactive Energy Strategy for Energy Trading of Proactive Distribution Company with Renewable Systems: a Robust/Stochastic Hybrid Technique, e-Prime-Advances in Electrical Engineering, Electron. Energy (2022) 100028.
[12] S. Janko, N.G. Johnson, Reputation-based competitive pricing negotiation and power trading for grid-connected microgrid networks, Appl. Energy 277 (2020) 115598.
[13] G.T. Udeh, S. Michailos, D. Ingham, K.J. Hughes, L. Ma, M. Pourkashanian, A modified rule-based energy management scheme for optimal operation of a hybrid PV-wind-Stirling engine integrated multi-carrier energy system, Appl. Energy 312 (2022) 118763.
[14] M. Daneshvar, S. Asadi, B. Mohammadi-Ivatloo, Energy trading possibilities in the modern multi-carrier energy networks, Grid Modernization—Future Energy Network Infrastructure, Springer, 2021, pp. 175–214.

[15] T. Pinto, R. Faia, M.A.F. Ghazvini, J. Soares, J.M. Corchado, Z. Vale, Decision support for small players negotiations under a transactive energy framework, IEEE Trans. Power Syst. 34 (5) (2018) 4015–4023.
[16] D. Forfia, M. Knight, R. Melton, The view from the top of the mountain: building a community of practice with the gridwise transactive energy framework, IEEE Power Energy Mag. 14 (3) (2016) 25–33.
[17] F. Rahimi, A. Ipakchi, F. Fletcher, The changing electrical landscape: end-to-end power system operation under the transactive energy paradigm, IEEE Power Energy Mag. 14 (3) (2016) 52–62.
[18] M. Daneshvar, B. Mohammadi-Ivatloo, K. Zare, S. Asadi, A. Anvari-Moghaddam, A novel operational model for interconnected microgrids participation in transactive energy market: a hybrid IGDT/stochastic approach, IEEE Trans. Ind. Inf. 17 (6) (2020) 4025–4035.
[19] M. Daneshvar, H. Eskandari, A.B. Sirous, R. Esmaeilzadeh, A novel techno-economic risk-averse strategy for optimal scheduling of renewable-based industrial microgrid, Sustain. Cities Soc. 70 (2021) 102879.
[20] M. Daneshvar, B. Mohammadi-ivatloo, S. Asadi, S. Galvani, Short term optimal hydro-thermal scheduling of the transmission system equipped with pumped storage in the competitive environment, Majlesi J. Electr. Eng. 14 (1) (2020) 77–84.
[21] R.H.M. Zargar, M.H.Y. Moghaddam, Development of a Markov-Chain-Based solar generation model for smart microgrid energy management system, IEEE Trans. Sustain. Energy 11 (2) (2019) 736–745.
[22] S.A. Janko, N.G. Johnson, Scalable multi-agent microgrid negotiations for a transactive energy market, Appl. Energy 229 (2018) 715–727.
[23] J.F. Martins, E. Romero-Cadaval, D. Vinnikov, M. Malinowski, Transactive Energy: power Electronics Challenges, IEEE Power Electron. Mag. 9 (1) (2022) 20–32.
[24] F. Rahimi, S. Mokhtari, Grid-Edge Blockchain-Based Transactive Energy Platform: design and implementation, IEEE Electrif. Mag. 9 (3) (2021) 46–54.

CHAPTER

7

Applications of transactive energy technology in the multi-carrier energy network modernization

7.1 Introduction

Today's energy grid witnesses faster changes in different layers from energy generation to distribution [1]. The need for such changes can be the first ambiguity that can depend on which perspective we look at it from. The ever increasing demand for energy is one of these viewpoints that affects all layers of the system and forces energy planers to relook the energy structure by applying the necessary changes. Indeed, this issue drives the system to increase the capacity of energy generation, transmission, and distribution. However, approaching each of the aforementioned sectors and successfully reaching to required changes have highlighted critical challenges for the system. The challenge in the energy generation sector mostly lies in the economical and environmental aspects of energy production [2]. Using fossil fuels for power production not only imposes extra costs but also leaves harmful effects on the environment. On the other hand, an unlimited volume of them is not available on the earth and their resources are being depleted day by day. Such issues have moved the traditional power production systems toward the substitution with renewable energy sources (RESs) that are clean and affordable in their outputs [3]. From another perspective, increasing the capacity of transmission lines and distribution systems exposes the system to higher power losses and more costs [4]. Moreover, this type of centralized generation and supply challenges the reliable and secure energy supply. Hence, the

Emerging Transactive Energy Technology for Future Modern Energy Networks
DOI: https://doi.org/10.1016/B978-0-323-91133-7.00006-5

decentralized energy generation units are suggested to address the above-mentioned challenges, especially in their renewable forms [5]. Herein, when we are going ahead, some other critical challenges have appeared that need innovative solutions. The power production by RESs directly depends on climate changes that input great uncertainties to the system operation [6]. How to tackle this challenge to effectively utilize RESs for power generation is the main question that highlights the necessity for the presence of transactive energy (TE) technology in the system's planning. In this regard, TE is intended as one of the main ways for supporting the renewable-penetrated grid to keep its balance between supply and demand sides [7]. It is recognized as a strong contender for establishing the system's stability via setting up the energy-sharing possibilities at the market level. In the power system, the TE application can be started from the transmission level where the independent system operator seeks to clear the energy market based on the two-sided auction by considering the received bids from consumers and offers from the energy producers. Indeed, it can support the energy grid for reliable operation even when it is incorporated with a high portion of RESs. On the other hand, TE can be employed for the energy management of the buildings and smart homes to allow for energy balance on the demand side. In other words, it can facilitate the distribution system operator's responsibilities in managing energy interactions in the customer sector. This chapter is aimed to clarify the applications of TE from the operation level of energy networks to the distribution level. Moreover, the TE applications in developing energy exchanging strategies as well as multi-energy systems are elaborated in this chapter.

7.2 TE application in the operation of energy networks

As defined in the previous chapter, TE is a set of control and economic mechanisms that is targeted to balance energy between supply and demand by taking different objectives into account as the key values [8]. Under this crucial mission, it is developed to conduct this by intending various aspects and from multifarious perspectives. Such developments have made it appropriate for different vital applications in the energy network. From one perspective, TE is engaged with cyber-physical systems to effectively detect malicious activities in continuously monitoring the system for secure operation. For example, the authors proposed a cyber-physical security model in [9] for covering TE control mechanisms in adopting autonomous decisions for demand response and energy generation aiming to prevent the vulnerable system from malicious attacks. In [10], the cyber-physical-society system is developed to combine the power distribution system, electricity market, and demand response for making a whole energy interaction ecosystem by using TE as a key idea for enabling the system to peer-to-peer (P2P) energy trading as well as value chain broadcasting. To characterize the prosumers' data given the mutual information theorem, the authors in [11] proposed a dynamic intelligent algorithm aiming to support the TE framework in the optimal operation of the system. Moreover, a discrete-time nonlinear autonomous dynamical system is developed in [12] to model the interactions among the total energy demand and real-time electricity price with the aim of assessing the impact of attacks on the system with the TE market framework. In [13], the authors have applied TE concepts to turn out a local energy market for energy sharing of fully renewable microgrids aiming to optimally operate the decarbonized power grid. A two-settlement TE market is developed in [14] to promote TE sharing for distributed energy sources (DESs) aggregators and the microgrid operator to operate the system by pursuing total energy cost minimization and satisfying the network's constraints. In [15], TE is used for developing a new operational framework for microgrid energy trading aiming to exploit the power grid based on the two-stage robust stochastic model. A distributionally robust optimization (DRO) algorithm is developed in [16] for solving an optimization problem formulated using the TE framework to minimize the operation cost. This is while the TE framework is presented in [17] for the coordinated energy management of microgrids by applying the DRO algorithm for uncertainty modeling. On the other hand, in [18], TE control is availed as the market-based control for establishing energy transactions among distributed microgrids with high renewables aiming to support the system to increase the utilization efficiency of RESs and smooth volatile production output. In addition, a new TE operational framework is suggested in [19] to enable aggregators to interact with the distribution network operator based on the negotiation process with the aim of ensuring grid security. This is while the TE control is applied in [20] for the chance-constrained scheduling of hybrid microgrids aiming to regulate energy interactions of the system. Another key application of TE systems is in the urban transportation networks. In this respect, different studies have exerted TE frameworks for improving energy interactions in the transportation sector. For instance, the authors in [21] built a multi-period traffic assignment model for characterizing the vehicular flows and scrutinizing multi-period optimal energy sharing for microgrids in the urban transportation network. In this work, an optimal multi-agent based energy exchanging scheme is presented based on the TE to facilitate P2P communication as well as privacy-preserving of the market participants. In [22], a hierarchical TE framework is developed to locally coordinate and reduce electric vehicle (EV) charging demand and DESs production in electric distribution grids. The mentioned study also presents two fully decentralized P2P trading models for energy sharing based on the modified alternating direction method of multipliers (ADMM).

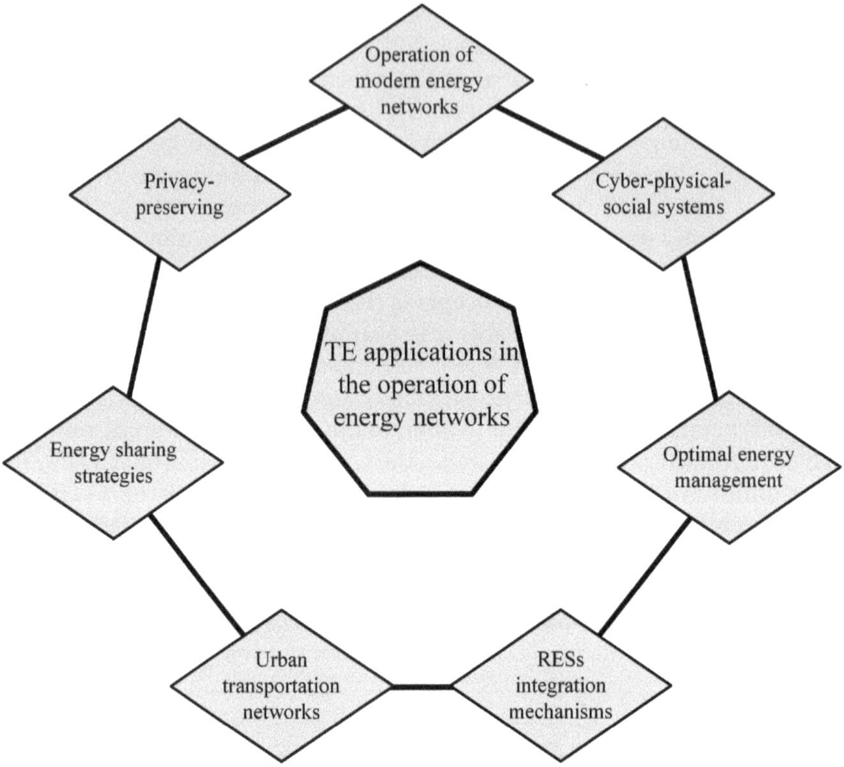

FIGURE 7.1 Some of the key applications of the TE in the operation of energy networks.

In another work [23], the authors used TE control for proposing a two-stage optimal charging scheme to manage real-time EV and day-ahead electricity procurement for minimizing overall operation costs. Moreover, a TE management system is proposed in [24] to control energy interactions of commercial parking lots with rooftop PV systems and EV charging units. Fig. 7.1 shows some of the key applications of the TE in the operation of energy networks.

7.3 TE application in consumer-side

Form another important overview, TE is one of the applicable technologies for modernizing the consumer side. Indeed, its application is not limited only to the operation of energy grids and widespread application is expected for the TE framework in the modern energy network. One of these critical applications is for the energy customer side of the modern grid where the smart buildings and homes are supposed to incorporate into the energy management programs. In recent years, several studies have concentrated on developing innovative solutions for the realization of the mentioned goal. In [25], a multi-agent framework is proposed based on the generalized hierarchical TE for coordinating the operation of various appliances in buildings as well as managing energy transacted at the cluster of buildings level. For energy management of smart homes in another work [26], the researchers developed a novel blockchain-based TE management system to enable them for performing vertical transactions with the aim of alleviating the grid load. In [27], an aggregation TE-based framework is proposed for prosumers to allow them to have privacy-preserved, proactive, and independent energy trading without depending on a real central entity. The authors presented a nested TE market methodology in [28] to effectively use the demand-side flexibility of residential consumers by participating in a local flexibility market based on the optimized transactive bidding. In [29], a hierarchical TE-based optimization framework is proposed to leverage and aggregate DESs flexibility for procuring contingency reserves in multiple local buildings. A TE market is used in [30] to determine optimal set points for plug-in electric vehicles by proposing a new energy management scheme for residential buildings. Low-carbon TE solutions are assessed in [31] to specify the interactive transaction behaviors of EVs and the positive roles of buildings in creating a sustainable TE community from human social space, data cyber space, and energy physical space. An energy management scheme is presented in [32] to enable various houses for adopting optimal decisions to effectively utilize their resources under the TE control. Reference [33] proposed a network constrained TE model for coordinating residential prosumers to satisfy households' preferences and providing a user-friendly strategy for them. Indeed, the TE is used not only for mitigating the problems of the distribution network in terms of voltage unbalance but also is developed to enable occupants in adjusting their willingness to pay/accept responsive assets. The TE technology is amended in [34] for the energy management of smart homes by upgrading the application of the home energy management system for involvement in the transactive energy interactions. The main goal of the TE energy management system is to balance the output of the energy generation units and energy consumption in the system. In [35], the TE management framework is presented to address cost optimization and grid overloading of the buildings. In the mentioned work, a multi-agent system architecture is also used for realizing the decentralized coordination for the energy management scheme aiming to preserve buildings' privacy as well as procure decision-making authority for them. Moreover, transactive control mechanisms are developed in [36] to allow buildings for energy interactions in an active distribution system and enable microgrids to cooperatively manage peak load. A novel decomposition algorithm is also used for designing a decentralized transactive control that makes preserving the detailed load information and profiles possible. Some of the prominent applications of the TE in the energy consumer side are denoted in Fig. 7.2.

7.4 TE application in energy trading strategies

One of the main missions of TE technology is to promote energy sharing strategies for renewable-based systems. In addition to the energy storage systems, demand-side energy management programs, DESs integration, and other flexibility improving methods, energy trading possibilities are recognized as another appropriate way of enhancing the grid's flexibility in the presence of numerous RESs. Similar advantages of energy trading methods alongside the ever-growing penetration of RESs have driven researchers to focus on developing energy trading strategies more than ever before. Thus, recent studies encompass a variety of energy trading strategies that facilitate the integration of RESs as one of the basic objectives of future modern energy networks. For example, in [37], a security-constrained decentralized P2P trading framework is proposed based on the TE technology to enable neighboring prosumers for direct energy trading with the aim of security and efficiency enhancement in the distribution system. A TE framework is proposed in [38] by applying auction theory and incorporating a system of agents to ease a TE market realization through an auctioneer. The proposed framework establishes a P2P energy trading community microgrid of residential buildings for possible monetary benefits. A distributed pricing strategy is presented in [39] for P2P-TE systems to manage voltage and line congestion as well as support the prosumers in maximizing their energy utilization via energy exchanging with neighbors under the arbitrator-free transactions. A P2P-TE trading framework is also investigated in [40] for the active participation of prosumers in the distribution network where the uncertainties are handled by proposing a vertex scenario-based robust optimization approach. In [41], the authors proposed an agent-based platform with a P2P energy transaction model based on the TE to intelligently manage energy in microgrids. A market mechanism is provided based on the P2P energy trading developed by the TE in [42] for optimally operating multiple energy hubs. The mentioned research applied a cooperative game theory for developing a convincing and fair payoff allocation program. An innovative multilevel TE optimization model is presented in [43] to explore the potential of blockchain technology in the scheduling of DESs within connected virtual power plants aiming to provide energy transactions among various peers. ATE-based P2P energy trading model is proposed in another work [44] for multi-microgrids to guarantee the security of the distribution power grid as well as improve the grid's flexibility by establishing energy trading among microgrids.

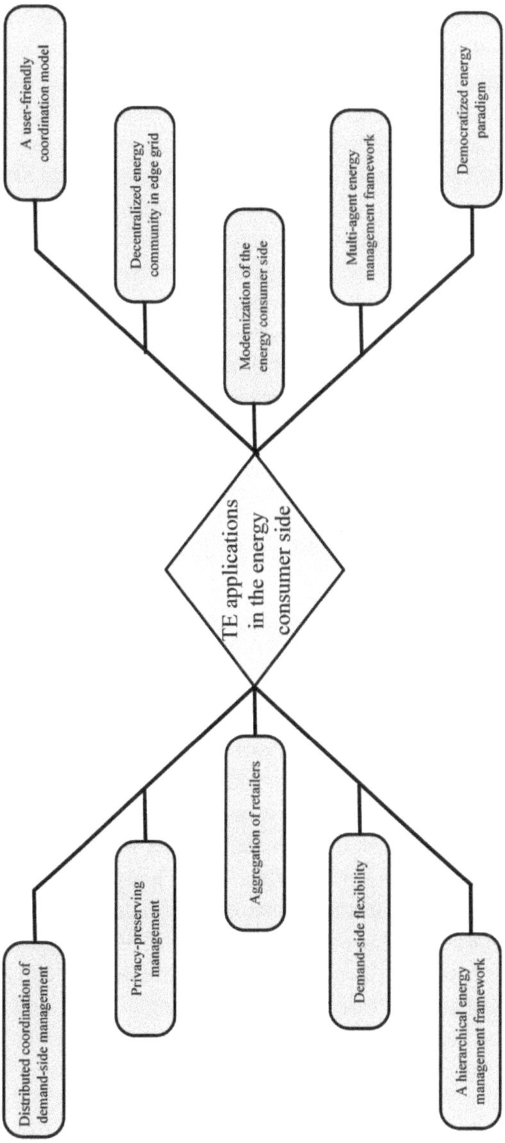

FIGURE 7.2 Some of the key applications of the TE in the energy consumer side.

In [45], a novel P2P transactive trading scheme is proposed given the multi-actor-attention-critic algorithm to maximize the flexibility value of prosumers and coordinately manage energy transactions within an energy community. A TE strategy is exerted in [46] for the optimal scheduling of a proactive distribution company equipped with RESs to determine the optimal set points for its energy trading under the robust/stochastic hybrid method for uncertainty modeling. In another work [47], a TE framework suggested for a distribution grid to enable consumers for involving in energy transactions by participating transactive agents in a local market for submitting their bids in the day-ahead planning. The authors in [48] employed TE technology for energy exchange control of microgrids by developing operational cases to scrutinize the microgrids' energy sharing with the aim of energy cost minimization. A comprehensive TE market framework is developed in [49] and incorporated with power grid layers and blockchain to offer ideal energy-sharing solutions for microgrids to ensure the security, transparency, and flexibility of the system. Moreover, an agent-based TE trading platform is proposed in [50] to integrate storage units into the energy management of microgrids aiming to create their participation conditions in the intra- and inter-microgrid TE markets. Fig. 7.3 portrays several key applications of the TE technology in designing different energy trading strategies.

7.5 TE application in multi-energy systems

Nowadays, hybrid energy systems have become prevalent units of grid development by substantial advancements in multi-energy systems. They effectively highlighted that independent energy networks are not more applicable for modern grid expansion. Indeed, realistic viewpoints for optimal operation of modern energy networks render the fact of interdependently exploiting energy grids due to the widespread presence of hybrid energy systems. Herein, multi-energy trading is known as the key possibility for such coupled energy infrastructure for the sustainability and flexibility of the system. In this regard, TE is widely applied in recent studies to make multi-energy trading possible for interconnected energy networks. For instance, the authors proposed a joint energy and reserve dispatch model for integrated heat and electricity units in [51] to coordinately manage energy transactions under the TE paradigm. Reference [52] proposed a bi-level TE trading framework to improve the operation efficiency of multi-carrier energy hubs in the optimal scheduling of the system with the social welfare maximization goal. In [53], a TE mechanism-supported energy exchanging strategy is developed for coordinating interconnected multi-energy complementary microgrids aiming to optimize multi-energy sharing in a regional coupled energy system. A distributionally

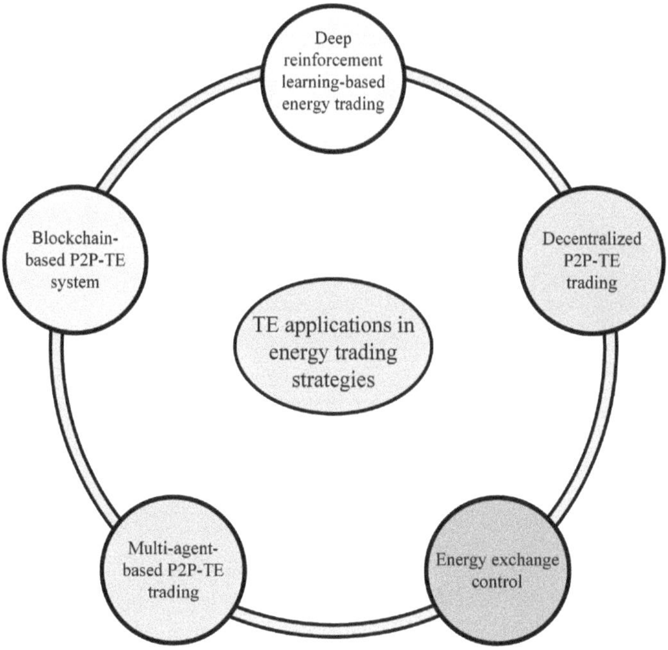

FIGURE 7.3 Several key applications of the TE technology in designing different energy trading strategies.

robust chance-constrained model is proposed based on the TE approach in [54] to manage energy and gas transactions in the optimal scheduling of coupled electrical and gas networks with the cost minimization objective. In [55], a transactive market design is presented for energy hubs to allow them to meet energy demand by participating in the TE market to adopt optimal energy transactions based on developing optimal bidding and offering strategies. In another work [56], a fully decentralized model is investigated to create the possibility of energy trading for multi-carrier energy hubs under the P2P energy trading framework within a TE market to supply energy loads in a least-cost paradigm. In [57], the interconnected microgrids are optimally scheduled using the potential of hydrogen energy storage units, in which the energy sharing is posed based on the proposed TE framework for reliably supplying energy in the network with 100 percent RESs. The authors in [58] used TE concepts to propose a new solution for fixing the energy dilemma in hub energies in a way the TE technology is deployed for energy exchanging in the power and water coupled network. In [59], the authors offered the TE-based solution for providing heat and electricity exchanging opportunities for energy hubs to encourage them to join energy interactions for meeting their energy

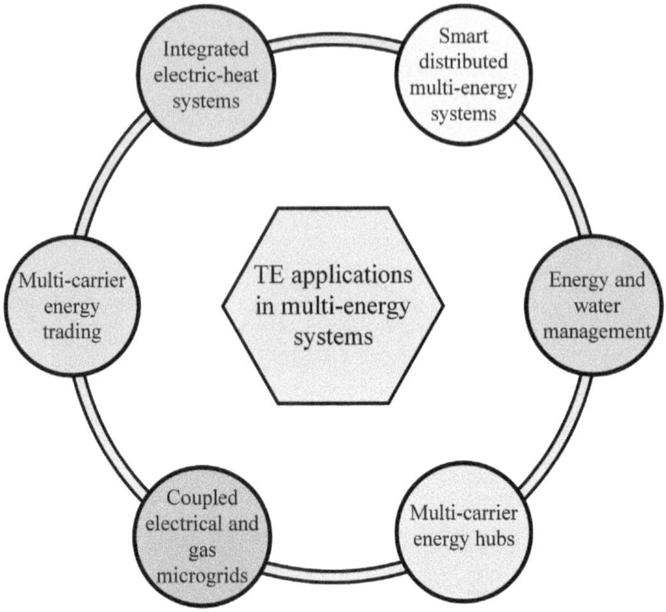

FIGURE 7.4 Some of the important applications of the TE technology for multi-energy systems.

demand in the modern interconnected heat and electricity network. In another research [60], energy interactions among the multi-carrier energy hubs are managed using the TE framework that provides the opportunity for multi-energy sharing with the aim of achieving affordable and clean energy in the hybrid grid.

The key operational models are developed using TE concepts for microgrid clusters in [61] to make electricity and thermal energy sharing possible for them and assess their energy interactions from the economic and technical stand of points. Additionally, a novel TE modeling and assessment framework is proposed in [62] for the technical and economic analysis of distributed multi-energy systems to provide flexible and cost-effective energy sharing. Some of the important applications of the TE technology for multi-energy systems are illustrated in Fig. 7.4.

7.6 Real-world implementations of TE systems

The extensive applications of the TE technology mentioned in previous sections indicate its potential for implementation in different real projects. Indeed, TE technology possesses clear and effective benefits over more common coordination systems, especially in terms of energy efficiency, scalability, and customer privacy. In Europe, the United States, and

Australia, research for the TE is mostly focused on power technology companies, regulators, energy supply companies, agent-based innovation in the field demonstrations, and household equipment. Some of the real-world implementations of TE systems are described as follows.

- **Olympic Peninsula Demonstration, 2006–07 [63]:** Partnered with several organizations on three extensive projects, the U.S. Department of Energy has concentrated on applying the TE systems for coordinating DESs with system operations. In this regard, the Olympic Peninsula of Washington state was the first area for the proof-of-concept TE implementation, in which a radial transmission connected to the Pacific Northwest power system feeds its electricity demand. The project is aimed at improving flexibility by coordinating DESs to remove or postpone the need for a transmission upgrade. Moreover, the project developed the TE application for achieving multiple objectives: municipal energy cost savings, wholesale price purchases by the commercial, residential, utility, distribution constraint, and system peak load management.
- **AEP Ohio grid SMART Real-Time Pricing Demonstration, 2010–14 [63]:** The AEP Ohio gridSMART project encompasses the SMART Choice as a real-time pricing component for creating a 5-min double-auction market to conduct the dispatch process of responsive loads. In this project, software agents are used for reflecting household occupants' preferences to develop an overall price flexibility curve for controlling various actions in the market system. The regional market operator is incorporated in the market clearing process using the related engine at the operations center for aggregating the bids received from households and forming a demand curve with the aim of calculating a supply bid and a clearing price. By broadcasting the clearing price to the households, the billing process is completed considering a tariff approved by the Public Utility Commission of Ohio as a regulator.
- **Pacific Northwest Smart Grid Demonstration (PNWSGD), 2010–15 [63]:** The PNWSGD consists of multiple states and cooperates with multiple utilities such as municipal and rural electric co-ops and investor-owned. The project sites encompass of distinct smart grid systems with different local objectives including demand responsiveness, improved efficiency, energy conservation, and improved reliability. The PNWSGD also employed TE technology for coordinating the operation of DESs and addressing regional goals, including flattening of system load and mitigation of RESs intermittency.
- **Transactive Home Energy Management Systems [64]:** In this project, the applicability of the TE is expanded for developing the home energy management system (HEMS) to create a dynamic energy balance on the demand side. To this end, the effectiveness of the TE-based HEMS is examined considering the energy management of smart homes in

Charlotte, North Carolina. A taxonomic GridLAB-D feeder model is adopted for modeling the relevant system. The obtained results proved the TE-based model can effectively synchronize the exploitation of air conditioners and peak-shaving.

- **TE Technology Implementations in Europe [63]:** PowerMatcher was a major European coordination mechanism that was developed based on TE technology. It is installed in several industrial sites and households to integrate multiple small producing and consuming electrical devices in the exploitation of the power infrastructure. Due to unbundling the European electricity sector into the network subsystem for operating the physical distribution and transmission networks as well as a market subsystem for the exchange of electricity commodities, the active distribution management along with the market operations are taken into account as the main application fields of the PowerMatcher. In this regard, the Couperus Smart Grid project (2011–15, The Netherlands) is one of the field experiences of the PowerMatcher technology.
- **TE Market for the Monash Microgrid [65]:** In the Monash Microgrid project in Australia, the TE paradigm is used for designing the renewable-penetrated microgrid. The TE concept is posed for managing energy interactions as well as establishing energy exchange in the Monash microgrid. On the other hand, as one of the main goals of this project, the TE is deployed for facilitating the integration of DESs in the system. This project is tested on the Monash Microgrid for analyzing the capability of the TE technology in the energy management of the system.

Fig. 7.5 demonstrates some of the prominent real-world implementations of TE systems.

7.7 Summary

As one of the basic requirements of grid modernization, this chapter scrutinizes the application of the new emerging TE technology in the multi-carrier energy network modernization. In this context, the application of the TE is discussed from different perspectives. The operation of energy networks is one of the critical fields of study that is benefited from the TE in developing multifarious models. However, the application of TE is not just restricted to the operation level of energy networks. It is applied to the energy customer-side to make end-users as active players in the energy interactions. Besides the aforementioned applications, one of the overarching goals for TE is to design innovative energy sharing strategies. Indeed, its application in forming various energy trading strategies is examined, especially in providing sufficient flexibility for RESs-based systems. By appearing multi-energy systems in the energy infrastructure, the idea of

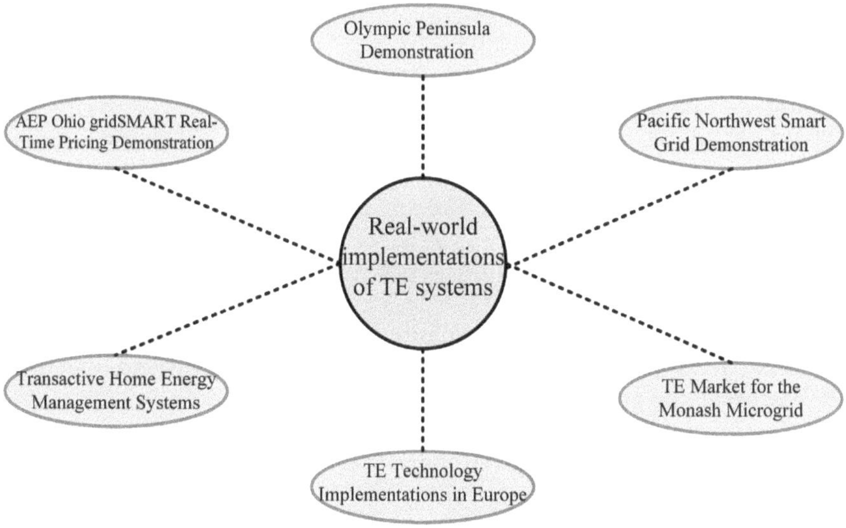

FIGURE 7.5 Some of the prominent real-world implementations of TE systems.

multi-energy trading is also raised. This issue has highlighted the need for capable technologies for establishing multi-energy trading in the system. Herein, the TE application is investigated in enabling multi-carrier energy networks for multi-energy trading in the hybrid network. Moreover, the real-world applications of the TE systems are discussed for the projects around the world that indicated its effective role in succeeding different practical projects. Analyzing different applications of the TE technology can support the system in effectively adopting and developing TE models for modernizing future multi-vector energy networks.

7.8 Questions

The careful analyzing whole of this chapter is expected to enable readers to effectively evaluate the following questions.

1) What are the necessities for TE technology in the modern energy grid?
2) How analyzing the applications of TE can be useful for future modern energy networks?
3) What are the key applications of the TE in the operation of energy networks?
4) What are the vital roles of the TE in operating future energy grids?
5) How TE can provide sufficient flexibility for the system with a high/full share of RESs?

6) How TE can support the energy customer-side to act as the active player in the energy interactions?
7) What are the applications of the TE in the energy management of the energy customer-side?
8) What are the key capabilities of the TE in developing innovative energy trading strategies?
9) How the TE can be applicable for designing new energy sharing strategies?
10) Why the multi-energy trading is essential for the flexibility of multi-carrier energy networks?
11) What are the TE applications in multi-energy systems?
12) What are the real-world implementations of TE systems and how they can affect the grid modernization process?
13) How evaluating different TE applications can assist the system to develop intelligent devices supporting the realization of the grid modernization schemes?

7.9 Suggestions

Given the topic of this chapter, the following suggestions are provided for the readers assisting them to pursue related research clues.

1) Developing TE models in accordance with the evolution of energy technologies can facilitate the optimal operation of future modern energy networks.
2) Applying the TE paradigm in the energy management of the consumer side can procure suitable conditions for end-users' participation in the energy interactions.
3) Advancing TE technology for designing novel energy trading strategies can support modern energy grids in maintaining sustainability and flexibility in the presence of RESs.
4) Extending the application of TE to multi-energy trading models can drive multi-carrier energy networks for adopting reliable multi-energy sharing strategies for the system's stability.

7.10 Future trends

Future trends can give a proper overview for the readers to form their related works and develop effective solutions for the remaining gaps, which are listed as follows for this chapter.

1) The holistic TE models need to be developed in line with the energy technology advancements to ease the exploitation of future modern energy networks.

2) The comprehensive TE paradigm is necessary for facilitating the involvement of end-users in the TE market aiming for purposeful energy management on the consumer side.
3) TE technology needs to be developed for designing new energy trading strategies to keep the grid's sustainability and flexibility in the high penetration of RESs.
4) The application of TE technology needs to be expanded to cover multi-energy trading for improving the multi-energy system's flexibility as well as remaining the stability of multi-carrier energy networks.

References

[1] S. An, H. Wang, X. Leng, Optimal operation of multi-micro energy grids under distribution network in Southwest China, Appl. Energy 309 (2022) 118461.

[2] M. Daneshvar, E. Babaei, Exchange market algorithm for multiple DG placement and sizing in a radial distribution system, J. Energy Manage. Technol. 2 (1) (2018) 54–65.

[3] M. Daneshvar, B. Mohammadi-ivatloo, S. Asadi, S. Galvani, Short term optimal hydro-thermal scheduling of the transmission system equipped with pumped storage in the competitive environment, Majlesi J. Electr. Eng. 14 (1) (2020) 77–84.

[4] M. Daneshvar, M. Abapour, B. Mohammadi-ivatloo, S. Asadi, Impact of optimal DG placement and sizing on power reliability and voltage profile of radial distribution networks, Majlesi J. Electr. Eng. 13 (2) (2019) 91–102.

[5] Y. Cao, et al., Development and transient performance analysis of a decentralized grid-connected smart energy system based on hybrid solar-geothermal resources; Techno-economic evaluation, Sustain. Cities Soc. 76 (2022) 103425.

[6] M. Daneshvar, H. Eskandari, A.B. S.irous, R. Esmaeilzadeh, A novel techno-economic risk-averse strategy for optimal scheduling of renewable-based industrial microgrid, Sustain. Cities Soc. 70 (2021) 102879.

[7] M. Daneshvar, B. Mohammadi-ivatloo, K. Zare, Integration of distributed energy resources under the transactive energy structure in the future smart distribution networks, in: Operation of Distributed Energy Resources in Smart Distribution Networks, Elsevier, 2018, pp. 349–379.

[8] D. Forfia, M. Knight, R. Melton, The view from the top of the mountain: building a community of practice with the gridwise transactive energy framework, IEEE Power Energy Mag. 14 (3) (2016) 25–33.

[9] Y. Zhang, et al., Cyber physical security analytics for transactive energy systems, IEEE Trans. Smart Grid 11 (2) (2019) 931–941.

[10] M. Song, Y. Cai, C. Gao, T. Chen, Y. Yao, H. Ming, Transactive energy in power distribution systems: paving the path towards cyber-physical-social system, Int. J. Electr. Power Energy Syst. 142 (2022) 108289.

[11] R. Zamani, M.P. M.oghaddam, M.-R. Haghifam, Dynamic Characteristics Preserving Data Compressing Algorithm For Transactive Energy Management Frameworks, IEEE Trans. Ind. Inf. (2022).

[12] K. Jhala, B. Natarajan, A. Pahwa, H. Wu, Stability of transactive energy market-based power distribution system under data integrity attack, IEEE Trans. Ind. Inf. 15 (10) (2019) 5541–5550.

[13] M. Daneshvar, B. Mohammadi-Ivatloo, K. Zare, S. Asadi, A. Anvari-Moghaddam, A novel operational model for interconnected microgrids participation in transactive energy market: a hybrid IGDT/stochastic approach, IEEE Trans. Ind. Inf. 17 (6) (2020) 4025–4035.

[14] B. Wang, C. Zhang, C. Li, G. Yang, Z.Y. D.ong, Transactive Energy Sharing in a Microgrid via an Enhanced Distributed Adaptive Robust Optimization Approach, IEEE Trans. Smart Grid 13 (3) (2022) 2279–2293.
[15] M. Daneshvar, B. Mohammadi-Ivatloo, K. Zare, S. Asadi, Two-stage robust stochastic model scheduling for transactive energy based renewable microgrids, IEEE Trans. Ind. Inf. 16 (11) (2020) 6857–6867.
[16] Y. Cao, et al., Optimal Energy Management for Multi-Microgrid Under a Transactive Energy Framework With Distributionally Robust Optimization, IEEE Trans. Smart Grid 13 (1) (2021) 599–612.
[17] Z. Liu, L. Wang, L. Ma, A transactive energy framework for coordinated energy management of networked microgrids with distributionally robust optimization, IEEE Trans. Power Syst. 35 (1) (2019) 395–404.
[18] W. Liu, J. Zhan, C. Chung, A novel transactive energy control mechanism for collaborative networked microgrids, IEEE Trans. Power Syst. 34 (3) (2018) 2048–2060.
[19] P. Hou, G. Yang, J. Hu, P.J. D.ouglass, Y. Xue, A Distributed Transactive Energy Mechanism for Integrating PV and Storage Prosumers in Market Operation, Engineering (2022).
[20] M. Daneshvar, B. Mohammadi-Ivatloo, K. Zare, M. Abapour, S. Asadi, A. Anvari-Moghaddam, Chance-constrained scheduling of hybrid microgrids under transactive energy control, Int. J. Energy Res. 45 (7) (2021) 10173–10190.
[21] Y. Liu, Y. Wang, Y. Li, H.B. G.ooi, H. Xin, Multi-agent based optimal scheduling and trading for multi-microgrids integrated with urban transportation networks, IEEE Trans. Power Syst. 36 (3) (2020) 2197–2210.
[22] J. Yang, T. Wiedmann, F. Luo, G. Yan, F. Wen, G.H. B.roadbent, A Fully Decentralized Hierarchical Transactive Energy Framework for Charging EVs with Local DERs in Power Distribution Systems, IEEE Transactions on Transportation Electrification (2022).
[23] Z. Liu, Q. Wu, K. Ma, M. Shahidehpour, Y. Xue, S. Huang, Two-stage optimal scheduling of electric vehicle charging based on transactive control, IEEE Trans. Smart Grid 10 (3) (2018) 2948–2958.
[24] A. Mohammad, R. Zamora, T.T. L.ie, Transactive Energy Management of PV-Based EV Integrated Parking Lots, IEEE Syst. J. 15 (4) (2020) 5674–5682.
[25] R. Chandra, S. Banerjee, K.K. R.adhakrishnan, S.K. P.anda, Transactive energy market framework for decentralized coordination of demand side management within a cluster of buildings, IEEE Trans. Ind. Appl. 57 (4) (2021) 3385–3395.
[26] Q. Yang, H. Wang, Privacy-preserving transactive energy management for IoT-aided smart homes via blockchain, IEEE Internet of Things Journal 8 (14) (2021) 11463–11475.
[27] Y. Xiao, X. Wang, P. Pinson, X. Wang, Transactive energy based aggregation of prosumers as a retailer, IEEE Trans. Smart Grid 11 (4) (2020) 3302–3312.
[28] M.S.H. Nizami, M. Hossain, K. Mahmud, A nested transactive energy market model to trade demand-side flexibility of residential consumers, IEEE Trans. Smart Grid 12 (1) (2020) 479–490.
[29] B. Hu, et al., A hierarchical transactive energy management framework for optimizing the reserve profile in district energy systems, CSEE J. Power Energy Syst. 7 (5) (2021) 922–931.
[30] H. Saber, H. Ranjbar, S. Fattaheian-Dehkordi, M. Moeini-Aghtaie, M. Ehsan, M. Shahidehpour, Transactive Energy Management of V2G-Capable Electric Vehicles in Residential Buildings: an MILP Approach, IEEE Trans. Sustain. Energy (2022).
[31] Y. Wu, Y. Wu, J.M. G.uerrero, J.C. V.asquez, Decentralized transactive energy community in edge grid with positive buildings and interactive electric vehicles, Int. J. Electr. Power Energy Syst. 135 (2022) 107510.
[32] M.N. A.kter, M.A. M.ahmud, M.E. H.aque, A.M. O.o, An optimal distributed energy management scheme for solving transactive energy sharing problems in residential microgrids, Appl. Energy 270 (2020) 115133.

[33] H. Saber, M. Ehsan, M. Moeini-Aghtaie, H. Ranjbar, M. Lehtonen, A User-Friendly Transactive Coordination Model for Residential Prosumers Considering Voltage Unbalance in Distribution Networks, IEEE Trans. Ind. Inf. (2022).
[34] M. Daneshvar, M. Pesaran, B. Mohammadi-ivatloo, Transactive energy in future smart homes, in: The Energy Internet, Elsevier, 2019, pp. 153–179.
[35] M.S.H. Nizami, M.J. H.ossain, E. Fernandez, Multiagent-based transactive energy management systems for residential buildings with distributed energy resources, IEEE Trans. Ind. Inf. 16 (3) (2019) 1836–1847.
[36] Y. Ge, H. Ye, K.A. L.oparo, Agent-based privacy preserving transactive control for managing peak power consumption, IEEE Trans. Smart Grid 11 (6) (2020) 4883–4890.
[37] L. Wang, et al., Security constrained decentralized peer-to-peer transactive energy trading in distribution systems, CSEE J. Power Energy Syst. 8 (1) (2021) 188–197.
[38] H.R. B.okkisam, S. Singh, R.M. A.charya, M. Selvan, Blockchain-based peer-to-peer transactive energy system for community microgrid with demand response management, CSEE J. Power Energy Syst. (2021).
[39] M.H. U.llah, J.-D. Park, Peer-to-peer energy trading in transactive markets considering physical network constraints, IEEE Trans. Smart Grid 12 (4) (2021) 3390–3403.
[40] X. Chang, Y. Xu, H. Sun, Vertex scenario-based robust peer-to-peer transactive energy trading in distribution networks, Int. J. Electr. Power Energy Syst. 138 (2022) 107903.
[41] L. Gomes, Z.A. V.ale, J.M. C.orchado, Multi-agent microgrid management system for single-board computers: a case study on peer-to-peer energy trading, IEEE Access 8 (2020) 64169–64183.
[42] W. Gan, M. Yan, W. Yao, J. Wen, Peer to peer transactive energy for multiple energy hub with the penetration of high-level renewable energy, Appl. Energy 295 (2021) 117027.
[43] M. Gough, et al., Blockchain-based Transactive Energy Framework for Connected Virtual Power Plants, IEEE Trans. Ind. Appl. (2021).
[44] M. Yan, M. Shahidehpour, A. Paaso, L. Zhang, A. Alabdulwahab, A. Abusorrah, Distribution network-constrained optimization of peer-to-peer transactive energy trading among multi-microgrids, IEEE Trans. Smart Grid 12 (2) (2020) 1033–1047.
[45] Y. Ye, Y. Tang, H. Wang, X.-P. Zhang, G. Strbac, A Scalable Privacy-Preserving Multi-Agent Deep Reinforcement Learning Approach for Large-Scale Peer-to-Peer Transactive Energy Trading, IEEE Trans. Smart Grid 12 (6) (2021) 5185–5200.
[46] M. Daneshvar, B. Mohammadi-Ivatloo, K. Zare, A. Anvari-Moghaddam, Transactive Energy Strategy for Energy Trading of Proactive Distribution Company with Renewable Systems: a Robust/Stochastic Hybrid Technique, e-Prime-Advances in Electrical Engineering, Electron. Energy (2022) 100028.
[47] A. Taghizadeh, M. Montazeri, H. Kebriaei, Deep Reinforcement Learning-Aided Bidding Strategies for Transactive Energy Market, IEEE Syst. J. (2022).
[48] M. Daneshvar, B. Mohammadi-Ivatloo, M. Abapour, S. Asadi, Energy exchange control in multiple microgrids with transactive energy management, J. Modern Power Syst. Clean Energy 8 (4) (2020) 719–726.
[49] M.R. H.amouda, M.E. N.assar, M. Salama, A novel energy trading framework using adapted blockchain technology, IEEE Trans. Smart Grid 12 (3) (2020) 2165–2175.
[50] H.K. N.unna, A. Sesetti, A.K. R.athore, S. Doolla, Multiagent-based energy trading platform for energy storage systems in distribution systems with interconnected microgrids, IEEE Trans. Ind. Appl. 56 (3) (2020) 3207–3217.
[51] W. Zhao, et al., Transactive Energy-Based Joint Optimization of Energy and Flexible Reserve for Integrated Electric-Heat Systems, IEEE Access 9 (2021) 14491–14503.
[52] Y. Wang, Z. Huang, Z. Li, X. Wu, LL. L.ai, F. Xu, Transactive energy trading in reconfigurable multi-carrier energy systems, J. Modern Power Syst. Clean Energy 8 (1) (2019) 67–76.

[53] Z. Yang, J. Hu, X. Ai, J. Wu, G. Yang, Transactive energy supported economic operation for multi-energy complementary microgrids, IEEE Trans. Smart Grid 12 (1) (2020) 4–17.
[54] M. Daneshvar, B. Mohammadi-Ivatloo, M. Abapour, S. Asadi, R. Khanjani, Distributionally robust chance-constrained transactive energy framework for coupled electrical and gas microgrids, IEEE Trans. Ind. Electron. 68 (1) (2020) 347–357.
[55] M. Alipour, M. Abapour, S. Tohidi, S.G. F.arkoush, S.-B. Rhee, Designing Transactive Market for Combined Heat and Power Management in Energy Hubs, IEEE Access 9 (2021) 31411–31419.
[56] M.S. J.avadi, A.E. N.ezhad, A.R. J.ordehi, M. Gough, S.F. S.antos, J.P. C.atalão, Transactive energy framework in multi-carrier energy hubs: a fully decentralized model, Energy 238 (2022) 121717.
[57] M. Daneshvar, B. Mohammadi-Ivatloo, K. Zare, S. Asadi, Transactive energy management for optimal scheduling of interconnected microgrids with hydrogen energy storage, Int. J. Hydrogen Energy 46 (30) (2021) 16267–16278.
[58] M.J.V. Pakdel, F. Sohrabi, B. Mohammadi-Ivatloo, Multi-objective optimization of energy and water management in networked hubs considering transactive energy, J. Cleaner Prod. 266 (2020) 121936.
[59] M. Daneshvar, B. Mohammadi-Ivatloo, K. Zare, A novel transactive energy trading model for modernizing energy hubs in the coupled heat and electricity network, J. Cleaner Prod. 344 (2022) 131024.
[60] M. Khorasany, A. Najafi-Ghalelou, R. Razzaghi, B. Mohammadi-Ivatloo, Transactive energy framework for optimal energy management of multi-carrier energy hubs under local electrical, thermal, and cooling market constraints, Int. J. Electr. Power Energy Syst. 129 (2021) 106803.
[61] M. Daneshvar, et al., Chance-constrained models for transactive energy management of interconnected microgrid clusters, J. Cleaner Prod. 271 (2020) 122177.
[62] N. Good, E.A.M. Ceseña, C. Heltorp, P. Mancarella, A transactive energy modelling and assessment framework for demand response business cases in smart distributed multi-energy systems, Energy 184 (2019) 165–179.
[63] K. Kok, S. Widergren, A society of devices: integrating intelligent distributed resources with transactive energy, IEEE Power Energy Mag. 14 (3) (2016) 34–45.
[64] A. Pratt, D. Krishnamurthy, M. Ruth, H. Wu, M. Lunacek, P. Vaynshenk, Transactive home energy management systems: the impact of their proliferation on the electric grid, IEEE Electrif. Mag. 4 (4) (2016) 8–14.
[65] M. Khorasany, D. Azuatalam, R. Glasgow, A. Liebman, R. Razzaghi, Transactive energy market for energy management in microgrids: the monash microgrid case study, Energies 13 (8) (2020) 2010.

CHAPTER

8

Optimizing and modeling the transactive energy technology in modern multi-carrier energy networks

8.1 Introduction

Recent innovations in energy technology development indicate the necessity for a new ecosystem of energy that is expected to be based on the information technology platform. This ecosystem is supposed to integrate energy entities and insets the new connotations for different domains of the system [1]. Under the mentioned ecosystem, the integration of energy suppliers, buyers, and markets is targeted for bringing extensive synergies for the platform providers and the market participants. However, by identifying the widespread contribution of multi-vector energy in the grid structure, the ecosystem of energy needs to be upgraded to cover multi-carrier energy systems instead of the single one. The interconnected operation of energy networks offers substantial advantages, especially in terms of flexibility and stability [2]. In this regard, the tri-generation systems such as combined cooling, heating, and power (CCHP) provide the opportunity of simultaneously producing several energy carriers for the system and offer efficient multi-energy in coupled heat and electricity networks (CHENs) [3]. As CHENs can bring special benefits to the reliability and sustainability of the system, the optimal energy management of such structures has become a challenge due to the presence of uncertainties and multiple hybrid energy devices. Given the effectiveness of energy

Emerging Transactive Energy Technology for Future Modern Energy Networks
DOI: https://doi.org/10.1016/B978-0-323-91133-7.00007-7

trading solutions for the energy management of such structures, this chapter suggests transactive energy (TE) based model for the optimal operation of CHENs.

In the context of energy management, recent works include different studies for the optimal operation of CHENs. In most of them, several models are formed considering the critical need for handling multi-energy problems in the smart grid. In this respect, reference [4] introduced a transient model for offering a cost- and energy-efficient solution for the power-to-heat integration in the electrically heated regenerator storage. In [5], a Nash-Cournot equilibrium model is established to provide strategic investment in heat and power markets to effectively address the expansion planning of CHENs in the hybrid energy area. Due to the importance of transiting toward the low-carbon inertia, the authors intended a solar-assisted carbon capture-based model in [6] for the operation of integrated combined heat and power plants to acquire the low-carbon benefits in supplying industrial and residential districts. The necessity for simultaneously considering the technical, economical, and environmental characteristics in the grid modeling has led to the development of the multi-objective optimization strategy in [7] for the coordinated dispatch of CHENs based on the convex quadratic program. This is while, in [8], the mentioned factors are also considered for programming multi-objective optimization to maximize the economic and environmental benefits by employing a coupled decision-making method for choosing the ideal scheme. The authors in [9] developed a Manta Ray Foraging Optimization as a meta-heuristic algorithm to solve the economic heat and power dispatch with the aim of minimizing the total fuel cost by taking the related operational restrictions in cogeneration energy systems. For the economic dispatch of the combined heat and power production in another research [10], the stochastic fractal search optimization method is used to deal with the non-linearity of the problem considering the total cost minimization as the objective and fulfilling operational constraints. In [11], the authors proposed a math formulation of the hierarchical dispatch technique to coordinately operate CHENs given the feasible region of boundary variables for promoting wind power accommodation. Reference [12] also developed a manta ray foraging MRF optimizer to solve economic dispatch in cogeneration of heat and power by pursuing total cost minimization objective and satisfying technical goals. Moreover, a decentralized heat-electricity coordination framework is proposed in [13] to effectively share a part of RESs accommodation benefits from a power grid to the district heating network.

The investigated studies in the previous paragraph offer various models for the energy management of the interconnected heat and power system. However, all of the suggested models bring concerns for the continuous energy supply when the grid is equipped with a high/full range of RESs. As energy trading methods are known as suitable ways of establishing balance in such systems, developing a complete energy sharing-based

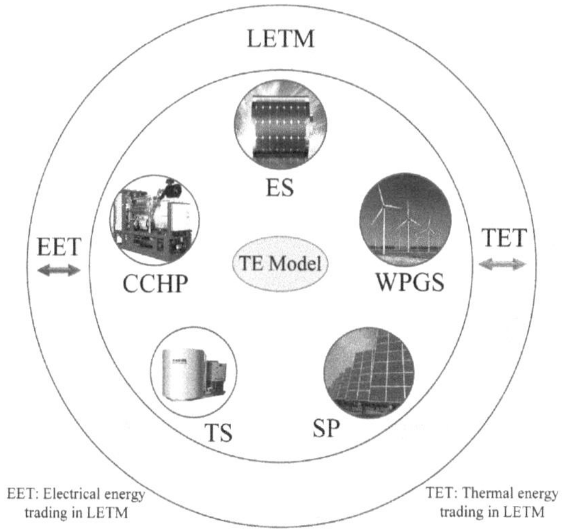

FIGURE 8.1 The schematic of microgrids in the CHEN.

model is ignored for the stability of the system in recent works. Due to this, the authors in [14] proposed a new model for modernizing energy hubs using the capability of TE technology in CHENs. Indeed, TE is used for manipulating a complete energy sharing-based model for enhancing the reliability of supply and facilitating the adoption of a high range of RESs in the modern energy grid. Given the important role of TE in easing the modernization of CHENs by offering energy sharing solutions, this chapter is aimed to deeply analyze the proposed model in [14] for highlighting how TE can be applied in designing energy sharing-based models as well as optimization problems. Therefore, in this chapter, the related problem formulation, simulation results, and discussions are scrutinized based on the model presented in [14].

8.2 Problem formulation

In this chapter, the main objective is the energy management of microgrids in CHENs considering the total cost minimization in the deregulated environment.

All microgrids encompass the CCHP unit, wind power generation system (WPGS), solar panels (SPs), electrical storage (ES), and thermal storage (TS) units. Fig. 8.1 shows the schematic of microgrids in the CHEN. Given this Fig., clean power production is intended by deploying WPGSs and SPs in the microgrid. The ESs and TSs are used for mitigating uncertainties effects and increasing the assurance of electrical and thermal energy supply. The CCHP unit is also operated to simultaneously deliver power, cooling, and heating energy to microgrids. The TE technology has provided the opportunity for free electrical and thermal energy sharing in the local energy trading market (LETM) to make microgrids more flexible than ever before. The energy sharing right with the main grid is also considered for microgrids for improving their cost-saving and easily satisfying technical goals.

The objective function for the optimal energy management of microgrids in CHEN is the minimization of total costs, which is formulated as follows:

$$\text{Min } OF_m = \sum_{s=1}^{S} \Gamma_s s \cdot \left[\begin{array}{l} \sum_{m=1}^{M}\sum_{t=1}^{T} (Fu_{m,t,s}^{PGS} \cdot p_t^{PGS} + Fu_{m,t,s}^{Bo} \cdot p_t^{Bo}) - \sum_{t=1}^{T} (P_{t,s}^{ET} \cdot p_t^{ET}) \\ -\sum_{i=1}^{B}\sum_{t=1}^{T} (P_{i,t}^{E,L} \cdot p_t^{SE} + (P_{i,t}^{H,L} + \sum_{m=1}^{M} (P_{m,t}^{HtDHN})) \cdot p_t^{SH} \end{array} \right] \tag{8.1}$$

where, OF_m is the objective function. Γ_s is the probability of scenarios. $Fu_{m,t,s}^{PGS}$ and $Fu_{m,t,s}^{Bo}$(p_t^{PGS} and p_t^{Bo}) are the consumed fuel (fuel price) by (of) the power generation system (PGS) and boiler in microgrid m, time t, and scenario s. $P_{t,s}^{ET}$ and p_t^{ET}($P_{i,t}^{E,L}$ and p_t^{SE}) are the shared energy (sold energy) and its price. $P_{i,t}^{H,L}$ is the heating energy demand. $P_{m,t}^{HtDHN}$ and p_t^{SH} are the shared heating energy with the heating energy grid and its price.

In this work, microgrids encounter uncertainties associated with RESs in power production that must be modeled for pragmatic modeling of the optimization problem. To capture uncertainties of microgrids, an autoregressive integrated moving average (ARIMA) is applied to deliver several scenarios related to the uncertainty deviations. Albeit the manifold scenarios are beneficial for the uncertainty modeling, a large number of them enter complexity for the relevant process. Due to this, the fast forward selection (FF) method is employed to decrease the volume of scenarios and select the most applicable of them for the optimization process. The math formulation of the ARIMA can be found in [15]. Furthermore, reference [16] provides sufficient information and a model for the FF.

To acquire the confident results from the optimization problem, considering the operational limitations of the engaged systems along with the network's constraints is essential. In this respect, the heat and power balance limitations are critical requirements for continuous energy supply that are given as:

$$\sum_{m=1}^{M} (P_{m,t}^{PGS} + P_{m,t}^{WPGS} + P_{m,t}^{SP} + P_{m,t}^{LtM} + \eta_m^{ESd} . P_{m,t}^{ES,D} - \frac{P_{m,t}^{ES,C}}{\eta_m^{ESc}} - P_{m,t}^{MtL} - P_{m,t}^{ET}) = \sum_{i=1}^{B} P_{i,t}^{E,L} \forall t \tag{8.2}$$

$$\sum_{m=1}^{M} (\eta_m^{TSd} . P_{m,t}^{TS,D} + HT_{m,t}^{CCHP}) = \sum_{m=1}^{M} (\frac{P_{m,t}^{TS,C}}{\eta_m^{TSc}} + P_{m,t}^{HtDHN}) + \sum_{i=1}^{B} P_{i,t}^{H,L} \forall t \tag{8.3}$$

where, $P_{m,t}^{PGS}$, $P_{m,t}^{WPGS}$, and $P_{m,t}^{SP}$ are the power outputs of the PGS, WPGS, and SPs. $P_{m,t}^{LtM}(P_{m,t}^{MtL})$ denotes the transmitted (received) power from the LETM to (by) the microgrid. $P_{m,t}^{ES,C}$ and $P_{m,t}^{ES,D}(\eta_m^{ESc}$ and $\eta_m^{ESd})$ are the charging and discharging power (efficiency) of the ES unit. $P_{i,t}^{E,L}$ states the power demand. $P_{m,t}^{TS,C}$ and $P_{m,t}^{TS,D}(\eta_m^{TSc}$ and $\eta_m^{TSd})$ are the charging and discharging thermal energy (efficiency) of the TS unit. $HT_{m,t}^{CCHP}$ is the produced heating energy by the CCHP. Eqs (8.2) and (8.3) model the power and heating energy balance constraints.

In the studied microgrids, WPGSs and SPs produce clean energy as RESs based on the following formulations.

$$\eta_m^{SP} \cdot SR_t^{SP} \cdot S_m^{SP} \geq P_{m,t}^{SP} \forall t, \forall m \tag{8.4}$$

$$P_{m,t}^{WPGS} = \begin{Bmatrix} 0 \upsilon_t^{Fo} < \upsilon_h^{C\cdots in}, \upsilon_t^{Fo} > \upsilon_h^{C\cdot out} \\ \bar{P}_m^{WPGS} \times \left(\dfrac{\upsilon_t^{Fo} - \upsilon_h^{C\cdot in}}{\upsilon_h^{Ra} - \upsilon_h^{C\cdots in}} \right)^3 \upsilon_h^{C\cdot in} \leq \upsilon_t^{Fo} \leq \upsilon_h^{Ra} \\ \bar{P}_m^{WPGS} \upsilon_h^{Ra} \leq \upsilon_t^{Fo} \leq \upsilon_h^{C\cdot out} \end{Bmatrix} \tag{8.5}$$

where, η_m^{SP} and S_m^{SP} indicate the efficiency and size of SPs. SR_t^{SP} and υ_t^{Fo} are the solar radiation and wind speed. $\bar{P}_m^{WPGS}$ is the rated power of the WPGS. The cut in $(\upsilon_h^{C.in})$, cut out $(\upsilon_h^{C.out})$, and rated (υ_m^{Ra}) speeds are the other parameters for modeling the WPGS.

The CCHP unit is used for generating heat and power to support electricity and heating energy networks. Its operational constraints are as follows:

$$Fp_{m,t}^{PGS} \leq S_m^{PGS} \cdot \gamma_{m,t}^{PGS} \forall t, \forall m \tag{8.6}$$

$$Fp_{m,t}^{Bo} \leq S_m^{Bo} \forall t, \forall m \tag{8.7}$$

$$\frac{(Fp_{m,t}^{PGS} - \gamma_{m,t}^{PGS}..x_m^{PGS})}{y_m^{PGS}} \geq P_{m,t}^{PGS} \forall t, \forall m \tag{8.8}$$

$$\eta_m^{Bo}.Fp_{m,t}^{Bo} + \eta_m^{PGS} PGS \cdot Fp_{m,t}^{PGS} \geq HT_{m,t}^{CCHP} \tag{8.9}$$

where, $S_m^{PGS}(S_m^{Bo})$ and $\gamma_{m,t}^{PGS}$ are the size of the PGS (boiler) and its ON/OFF status. x_m^{PGS} and y_m^{PGS} are the coefficients of modeling the PGS. η_m^{Bo} and η_m^{PGS} are the efficiency of the boiler and PGS.

All microgrids use TS for storing thermal energy and use the stored energy for balancing the district heating network during the day. The TS unit is operated subject to the following constraints.

$$\gamma_{m,t}^{TS,D} + \gamma_{m,t}^{TS,C} \leq 1 \forall t, \forall m \tag{8.10}$$

$$S_m^{TS} \geq P_{m,t}^{TS} \geq \chi_{\min}^{TS} \cdot S_m^{TS} \forall t, \forall m \tag{8.11}$$

$$P_{m,t}^{TS} = TS_m^{In} + \Delta t \cdot (P_{m,t}^{TS,C} - P_{m,t}^{TS,D}) \forall t = 1, \forall m \tag{8.12}$$

$$P_{m,t}^{TS} - P_{m,t-1}^{TS} = \Delta t \cdot (P_{m,t}^{TS,C} - P_{m,t}^{TS,D}) \forall t \geq 2, \forall m \tag{8.13}$$

$$\gamma_{m,t}^{TS,C} \cdot S_m^{TS} \cdot \chi_{\min}^{TS,C} \leq P_{m,t}^{TS,C} \leq \gamma_{m,t}^{TS,C} \cdot S_m^{TS} \cdot \chi_{\max}^{TS,C} \tag{8.14}$$

$$\gamma_{m,t}^{TS,D} \cdot S_m^{TS} \cdot \chi_{\min}^{TS,D} \leq P_{m,t}^{TS,D} \leq \gamma_{m,t}^{TS,D} \cdot S_m^{TS} \cdot \chi_{\max}^{TS,D} \tag{8.15}$$

where, $\gamma_{m,t}^{TS,D}$ and $\gamma_{m,t}^{TS,C}$ are the discharging and charging states for the TS. S_m^{TS} and TS_m^{In} are the size of TS and its initial stored energy. $\chi_{\min}^{TS,C}$ and $\chi_{\max}^{TS,C}(\chi_{\min}^{TS,D}$ and $\chi_{\max}^{TS,D})$ are the minimum and maximum charging (discharging) coefficients for the TS. Eq. (8.10) models the state of the TS while its stored energy constraint is indicated in (8.11). Eqs. (8.12) and (8.13) are for balancing thermal energy in the TS. Eqs (8.14) and (8.15) are the limitations for thermal energy charging and discharging.

In addition to the TS, ES is also deployed for alleviating the effects of RESs uncertainties and upsurging the energy supply confidence. The constraints for the ES are given as:

$$\gamma_{m,t}^{ES,D} + \gamma_{m,t}^{ES,C} \leq 1 \forall t, \forall m \tag{8.16}$$

$$S_m^{ES} \geq P_{m,t}^{ES} \geq \chi_{\min}^{ES} \cdot S_m^{ES} \forall t, \forall m \tag{8.17}$$

$$P_{m,t}^{ES} = ES_m^{In} + \Delta t \cdot (P_{m,t}^{ES,C} - P_{m,t}^{ES,D}) \forall t = 1, \forall m \tag{8.18}$$

$$P_{m,t}^{ES} - P_{m,t-1}^{ES} = \Delta t \cdot (P_{m,t}^{ES,C} - P_{m,t}^{ES,D}) \forall t \geq 2, \forall m \tag{8.19}$$

$$\gamma_{m,t}^{ES,C} \cdot S_m^{ES} \cdot \chi_{\min}^{ES,C} \leq P_{m,t}^{ES,C} \leq \gamma_{m,t}^{ES,C} . S_m^{ES} \cdot \chi_{\max}^{ES,C} \tag{8.20}$$

$$\gamma_{m,t}^{ES,D} \cdot S_m^{ES} \cdot \chi_{\min}^{ES,D} \leq P_{m,t}^{ES,D} \leq \gamma_{m,t}^{ES,D} \cdot S_m^{ES} \cdot \chi_{\max}^{ES,D} \tag{8.21}$$

where, $\gamma_{m,t}^{ES,D}$ and $\gamma_{m,t}^{ES,C}$ are the discharging and charging states for the ES. S_m^{ES} and ES_m^{In} are the size of ES and its initial stored energy. $\chi_{\min}^{ES,C}$ and $\chi_{\max}^{ES,C}(\chi_{\min}^{ES,D}$ and $\chi_{\max}^{ES,D})$ are the minimum and maximum charging (discharging) coefficients for the ES. Eq. (8.16) models the state of the ES while its stored energy constraint is denoted in (17). Eqs. (8.18) and (8.19) are for balancing electrical energy in the ES. Eqs. (8.20) and (8.21) are the limitations for electrical energy charging and discharging.

In this study, the constraints of the electric power distribution and district heating networks are intended for modeling CHEN in the simulation process. The AC power flow formulations and the related descriptions can be found in [17]. Moreover, the constant-flow variable-temperature mode is posed for modeling the district heating network, which its math formulas can be reached in [18].

8.3 Transactive energy model

Energy trading opportunities created for microgrids in this study are developed using the TE framework. The TE is recently applied as a reliable and secure method of balancing energy in renewable-based systems [19]. Its definition and attributes are fully elaborated in the previous chapters that indicate its capability in building a community of practice for market participants. In this research, the LETM is structured to be a suitable environment for energy sharing of microgrids in the local area. This new energy sharing environment is built under the TE paradigm that offers a free energy sharing possibility for microgrids by giving value to their objectives. Microgrids trade energy with each other in the LETM based on the following constraints:

$$\gamma_{m,t}^{O,E} + \gamma_{m,t}^{I,E} \leq 1 \forall m, \forall t \tag{8.22}$$

$$\gamma_{m,t}^{O,H} + \gamma_{m,t}^{I,H} \leq 1 \forall m, \forall t \tag{8.23}$$

$$Z^{E}.\gamma_{m,t}^{I,E} \geq P_{m,t}^{LtM} \forall m, \forall t \tag{8.24}$$

$$Z^{E}.\gamma_{m,t}^{O,E} \geq P_{m,t}^{MtL} \forall m, \forall t \tag{8.25}$$

$$Z^{H}.\gamma_{m,t}^{I,H} \geq H_{m,t}^{LtM} \forall m, \forall t \tag{8.26}$$

$$Z^{H}.\gamma_{m,t}^{O,H} \geq H_{m,t}^{MtL} \forall m, \forall t \tag{8.27}$$

$$\sum_{t} P_{m,t}^{LtM} = \sum_{t} P_{m,t}^{MtL} \forall m \tag{8.28}$$

$$\sum_{m} P_{m,t}^{LtM} = \sum_{m} P_{m,t}^{MtL} \forall t \tag{8.29}$$

$$\sum_{t} H_{m,t}^{LtM} = \sum_{t} H_{m,t}^{MtL} \forall m \tag{8.30}$$

$$\sum_{m} H_{m,t}^{LtM} = \sum_{m} H_{m,t}^{MtL} \forall t \tag{8.31}$$

where, $\gamma_{m,t}^{O,H}$ and $\gamma_{m,t}^{I,H}$ ($\gamma_{m,t}^{O,E}$ and $\gamma_{m,t}^{I,E}$) are the out putted and inputted heat (power) states in the LETM. Z^H and Z^E are upper coefficients for the heat and power trading in the LETM. Eqs. (8.22) and (8.23) denote the states of power and heating energy trading. Eqs. (8.24) and (8.25) ((8.26) and (8.27)) limit power (heating energy) transactions in the LETM. Eqs. (8.28) and (8.29) ((8.30) and (8.31)) are the power (heating energy) balance constraints in the LETM.

8.4 Simulation results

In this study, the chapter targets the evaluation of TE effectiveness in the scheduling of microgrids in the CHEN. For this aim, the combined version of the 10-node district heating [20] and IEEE 10-bus electricity

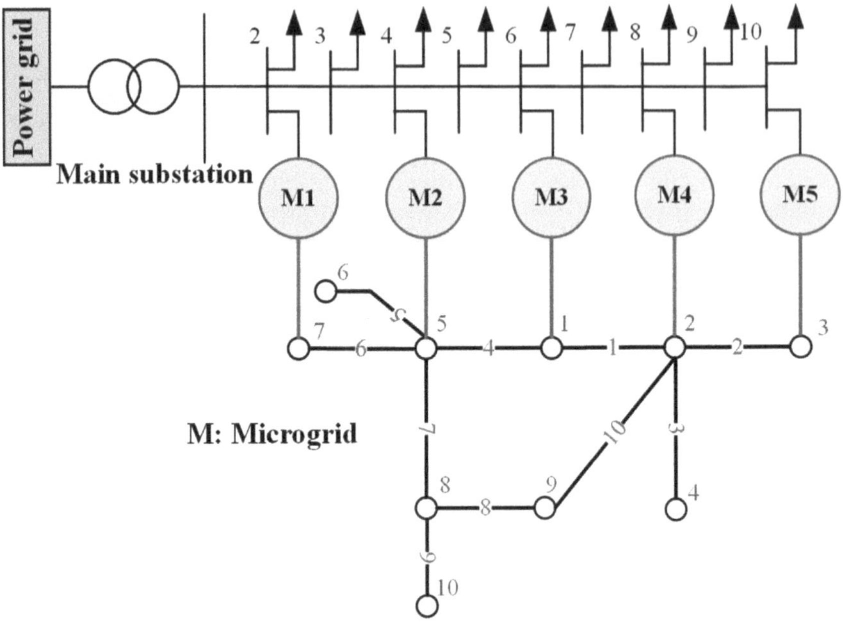

FIGURE 8.2 The schematic of the hybrid studied test system.

[21] grids is considered to assess the optimal energy management problem. The schematic of the hybrid studied test system is illustrated in Fig. 8.2.

The studied microgrid is equipped with the CCHP as a controllable unit for generating power and heat [22], the WPGS and SP for carbon-free energy production [23], the ES for alleviating uncertainties effects of RESs [22], and the TS for ensuring thermal energy supply [24]. In this research, two case studies are intended for evaluating the TE applicability. The first one is without uncertainty modeling while the second one models uncertainties in the system. The operation costs for both cases I and II are illustrated in Fig. 8.3.

According to Fig. 8.3, five microgrids are intended for assessing the optimal energy management in the CHEN. The results in this Fig. denote that all microgrids have lower total operation costs in Case I without uncertainty modeling. However, all of them have gained more costs in Case II due to considering the realistic states in the uncertainty modeling process. In this respect, Fig. 8.4 illustrates the outputs of the WPGS, ES, and the CCHP during a day.

In Fig. 8.4, from 1 am to 8 am, the higher power production by the WPGS has led to providing suitable conditions for meeting energy and the system did not need the discharging of ES in the high range as well as the maximum operation of the CCHP. However, from 10 am to 4 pm,

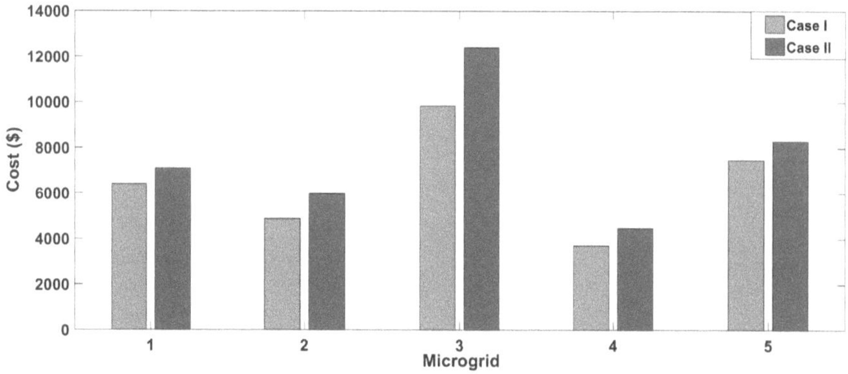

FIGURE 8.3 Operation costs for both cases I and II.

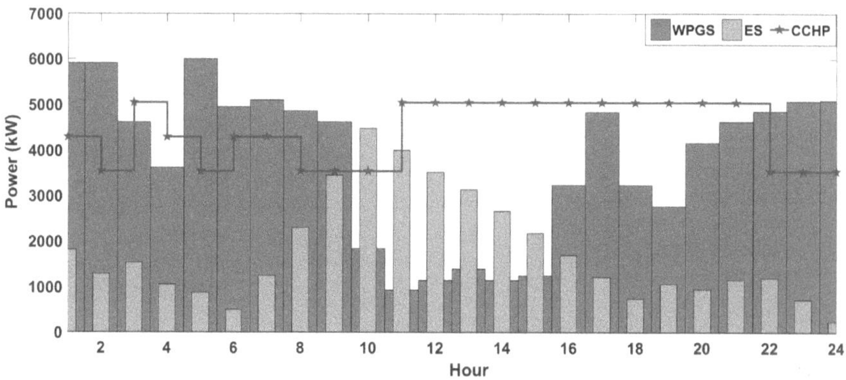

FIGURE 8.4 The outputs of the WPGS, ES, and the CCHP during a day.

the substantial depletion in the WPGS output has resulted in a significant increment in the discharging of the ES and working the CCHP at the maximum level. This is while by increasing the WPGS production from 4 pm, the ES discharging is decreased and the CCHP level is also reduced in the three last hours of the day. In this regard, Fig. 8.5 demonstrates the output of SPs and energy sharing with the power grid and in the LETM.

In Fig. 8.5, energy is purchased from the main grid from 1 am to 6 am when the energy price is low for meeting the energy demand of the mentioned period of time. However, from 7 am, by increasing the output of SPs, a portion of extra energy is delivered to the power grid and the energy sharing in the LETM is used for balancing energy at peak times. By decreasing the solar power production to zero at 8 pm, the energy trading with the main grid and in the LETM are considered for supporting microgrids to meet energy load. In the optimal energy management of CHEN, thermal energy is another carrier of energy that is evaluated in energy interactions. Fig. 8.6 shows the changes in the TS output and

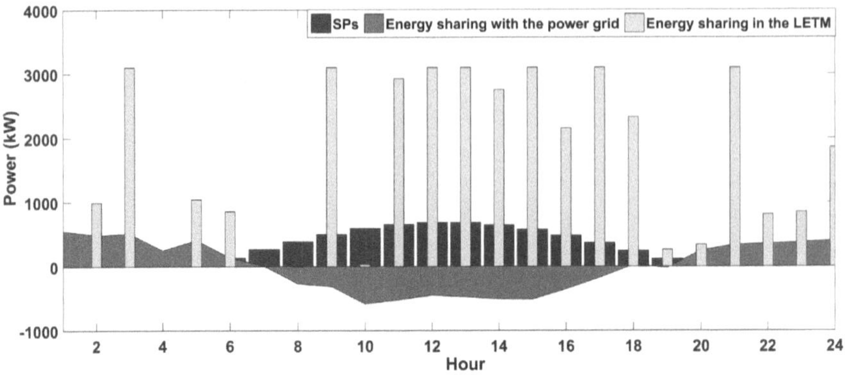

FIGURE 8.5 The output of SPs and energy sharing with the power grid and in the LETM.

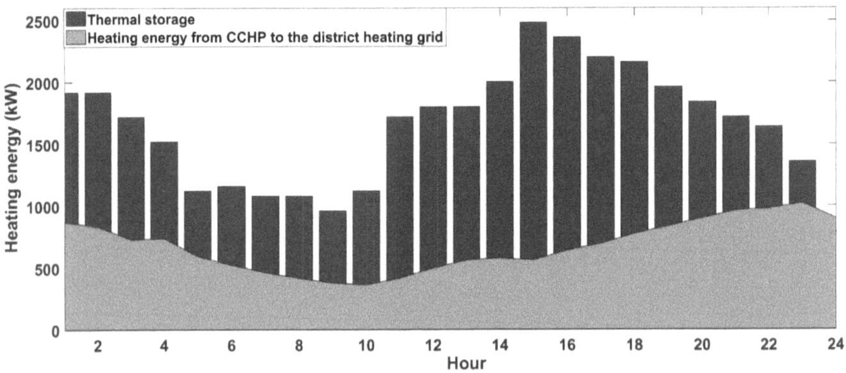

FIGURE 8.6 Changes in the TS output and delivered heating energy from the CCHP to the district heating network.

delivered heating energy from the CCHP to the district heating network.

Given Fig. 8.6, in the morning and evening hours, the more thermal energy demand has driven the optimal scheduling of microgrids to rely on more discharging TS and operating the CCHP unit to continuously meet the thermal energy load of the mentioned period of time. However, at the mid-hours of the day, the system faces lower TS discharging and CCHP generation due to the lower thermal energy consumption at peak hours. In this respect, the TS is used as the supportive device for the CCHP unit in dynamically balancing thermal energy supply and demand.

8.5 Summary

Given the necessity for developing practical models for integrated energy systems with a high penetration of RESs, this chapter proposes a new TE-based model to optimally manage power and heat in microgrids of the CHEN. The studied system encompasses five microgrids with the CCHP, ES, TS, SPs, and WPGS to analyze the effectiveness of the proposed model in the hybrid network. As the grid faces uncertainties of the different stochastic sources, uncertainty modeling was essential for the accurate optimization model. Thus, it is done by applying ARIMA for scenario generation and FF for scenario reduction. In this study, the combined version of the 10-node district heating and IEEE 10-bus electricity grids is selected to validate the proposed model. The results proved the applicability of the TE in providing energy sharing opportunities for microgrids to trade energy with each other in the LETM. Indeed, energy sharing in the LETM increased the flexibility of microgrids in meeting energy demand and assisted the system to keep its sustainability in the presence of RESs.

8.6 Questions

The careful analyzing whole of this chapter is expected to enable readers to effectively evaluate the following questions:

1) How TE technology can be modeled and optimized in modern multi-carrier energy networks?
2) How multi-energy can be optimally managed in the hybrid system with a high share of RESs?
3) What is the key role of TE in the optimal energy management of the integrated energy system?
4) How TE can improve the flexibility of microgrids in the energy supply process?
5) What is the role of LETM in establishing energy sharing opportunities for multi-microgrids?
6) What are the effective guidelines for energy sharing among microgrids in the LETM under the TE paradigm?
7) How different controllable units, storage systems, and RESs can be mathematically modeled in the optimization problem?
8) How TE can be mathematically implemented for energy sharing of microgrids in the LETM?
9) What are the results of applying TE technology for the energy management of the CHEN?
10) How TE can improve the realization of the technical and economic goals of microgrids in the deregulated environment?
11) What are the advantages of modeling energy networks integrately instead of independently?

8.7 Suggestions

Given the topic of this chapter, the following suggestions are provided for the readers assisting them to pursue related research clues:

1) Developing new models based on the TE technology can create effective energy sharing remedies for energy balancing challenges of future modern multi-energy networks with a high share of RESs.
2) Structuring local energy markets for energy sharing of different players can support the grid in mostly reducing dependencies on the main grid with fossil fuels and expanding the transmission lines with power losses.
3) Designing TE-based architecture can enable the integrated energy network to flexibly act in energy supply and keep the sustainability of the energy infrastructure.
4) Advancing energy sharing mechanisms can be useful strategies for the optimal energy management of future modern multi-vector energy networks.

8.8 Future trends

Future trends can give a proper overview for the readers to form their related works and develop effective solutions for the remaining gaps, which are listed as follows for this chapter:

1) Local energy markets are the necessary area for energy sharing of various market participants that offer tremendous benefits in terms of reducing the dependencies of microgrids to the main grid for energy supply as well as decreasing the need for new transmission lines or expanding the existent ones with power losses.
2) TE technology is one of the confident ways of dealing with the energy balance challenge of the modern multi-carrier energy network with a high penetration of renewable units that its development is the critical requirement of hybrid grids.
3) Different energy sharing mechanisms need to be advanced for the optimal energy management of the multi-vector energy grid to effectively allow the system for sustainability and stability.
4) Energy sharing architectures need to be developed based on the capable technologies that promote the interoperability in the integrated energy network and also support the energy management schemes to make the energy structure more flexible and sustainable.

References

[1] R. Masiello, J.R. Aguero, Sharing the ride of power: understanding transactive energy in the ecosystem of energy economics, IEEE Power Energy Mag. 14 (3) (2016) 70–78.

[2] N. Liu, J. Wang, L. Wang, Distributed energy management for interconnected operation of combined heat and power-based microgrids with demand response, J. Modern Power Syst. Clean Energy 5 (3) (2017) 478–488.

[3] Q. Wang, J. Liu, Y. Hu, X. Zhang, Optimal operation strategy of multi-energy complementary distributed CCHP system and its application on commercial building, IEEE Access 7 (2019) 127839–127849.

[4] S. Belik, V. Dreissigacker, S. Zunft, Power-to-heat integration in regenerator storage: enhancing thermal storage capacity and performance, J. Energy Storage 50 (2022) 104570.

[5] S. Chen, Z. Wei, G. Sun, Z. Yizhou, Strategic Investment in Power and Heat Markets: a Nash-Cournot Equilibrium Model, IEEE Trans. Ind. Inf. (2022).

[6] X. Guo, S. Lou, Y. Wu, Y. Wang, Low-Carbon Operation of Combined Heat and Power Integrated Plants Based on Solar-Assisted Carbon Capture, J. Modern Power Syst. Clean Energy (2021).

[7] X. Liu, F. Zhang, Q. Sun, W. Zhong, Multi-objective optimization strategy of integrated electric-heat system based on energy storage situation division, IEEE Access 9 (2021) 19004–19024.

[8] Y. Chen, X. Hu, W. Xu, Q. Xu, J. Wang, P.D. Lund, Multi-objective optimization of a solar-driven trigeneration system considering power-to-heat storage and carbon tax, Energy 250 (2022) 123756.

[9] A.M. Shaheen, A.R. Ginidi, R.A. El-Sehiemy, S.S. Ghoneim, Economic power and heat dispatch in cogeneration energy systems using manta ray foraging optimizer, IEEE Access 8 (2020) 208281–208295.

[10] M.I. Alomoush, Optimal combined heat and power economic dispatch using stochastic fractal search algorithm, J. Modern Power Syst. Clean Energy 8 (2) (2020) 276–286.

[11] T. Jiang, et al., Hierarchical dispatch method for integrated heat and power systems based on a feasible region of boundary variables, CSEE J. Power Energy Syst. 6 (3) (2020) 543–553.

[12] A.M. Shaheen, A.R. Ginidi, R.A. El-Sehiemy, E.E. Elattar, Optimal economic power and heat dispatch in Cogeneration Systems including wind power, Energy 225 (2021) 120263.

[13] J. Yang, A. Botterud, N. Zhang, Y. Lu, C. Kang, A cost-sharing approach for decentralized Electricity–Heat operation with renewables, IEEE Trans. Sustain. Energy 11 (3) (2019) 1838–1847.

[14] M. Daneshvar, B. Mohammadi-Ivatloo, K. Zare, A novel transactive energy trading model for modernizing energy hubs in the coupled heat and electricity network, J. Clean Prod. 344 (2022) 131024.

[15] M. Daneshvar, B. Mohammadi-Ivatloo, K. Zare, S. Asadi, Transactive energy management for optimal scheduling of interconnected microgrids with hydrogen energy storage, Int. J. Hydrogen Energy 46 (30) (2021) 16267–16278.

[16] M. Daneshvar, B. Mohammadi-ivatloo, S. Asadi, S. Galvani, Short term optimal hydro-thermal scheduling of the transmission system equipped with pumped storage in the competitive environment, Majlesi J. Electr. Eng. 14 (1) (2020) 77–84.

[17] M. Daneshvar, E. Babaei, Exchange market algorithm for multiple DG placement and sizing in a radial distribution system, J. Energy Manage. Technol. 2 (1) (2018) 54–65.

[18] Y. Chen, W. Wei, F. Liu, E.E. Sauma, S. Mei, Energy trading and market equilibrium in integrated heat-power distribution systems, IEEE Trans. Smart Grid 10 (4) (2018) 4080–4094.
[19] M. Daneshvar, B. Mohammadi-ivatloo, K. Zare, Integration of distributed energy resources under the transactive energy structure in the future smart distribution networks, in: Operation of Distributed Energy Resources in Smart Distribution Networks, Elsevier, 2018, pp. 349–379.
[20] Y. Cao, W. Wei, L. Wu, S. Mei, M. Shahidehpour, Z. Li, Decentralized operation of interdependent power distribution network and district heating network: a market-driven approach, IEEE Trans. Smart Grid 10 (5) (2018) 5374–5385.
[21] M. Daneshvar, B. Mohammadi-Ivatloo, K. Zare, M. Abapour, S. Asadi, A. Anvari-Moghaddam, Chance-constrained scheduling of hybrid microgrids under transactive energy control, Int. J. Energy Res. 45 (7) (2021) 10173–10190.
[22] Y. Chen, M. Hu, Balancing collective and individual interests in transactive energy management of interconnected micro-grid clusters, Energy 109 (2016) 1075–1085.
[23] M. Daneshvar, H. Eskandari, A.B. Sirous, R. Esmaeilzadeh, A novel techno-economic risk-averse strategy for optimal scheduling of renewable-based industrial microgrid, Sustain. Cities Soc. 70 (2021) 102879.
[24] M. Daneshvar, M. Pesaran, B. Mohammadi-ivatloo, Transactive energy integration in future smart rural network electrification, J Clean Prod 190 (2018) 645–654.

CHAPTER

9

A case study in the hybrid energy network with 100 percent renewable energy resources and future directions

9.1 Introduction

Over the recent years, the climate crisis has driven energy systems to use clean energy units, particularly renewable energy sources (RESs) for energy production [1]. The harmful effects of traditional units have led to the aim of transiting to the economies predominantly powered by RESs [2]. This transition to a carbon-free economy needs to be undergone in different parts of the system from primary energy production systems to the end consumers [3]. In this respect, as fossil fuel-based distributed generators offer controllable energy supply, their greenhouse gas emissions and fuel costs have created considerable concerns for network expansion [4]. Due to presenting certain advantages, especially in terms of environmental and economic, future modern energy networks (FMENs) have the main target of using 100 percent RESs for producing energy to approach the realization of decarbonization goals. On the other hand of presenting considerable benefits by RESs, their natural nature brings specific barriers to the widespread usage of them [5]. This issue also affects the reliability of the power supply as the critical factor in the distribution network [4]. Indeed, their uncertainties in energy production hinder the accurate planning for the energy grid and are the main challenge for their presence in the energy landscape. What are the appropriate strategies to cope with intermittences

Emerging Transactive Energy Technology for Future Modern Energy Networks

DOI: https://doi.org/10.1016/B978-0-323-91133-7.00004-1

of RESs require the energy systems to apply practical solutions. However, such solutions need to be empowered by capable technologies for making their implementation in practice feasible. Energy trading is one of the promising ways of promoting RESs in the energy generation sector. This is while hybrid networks require innovative energy sharing strategies for adopting 100 percent RESs. Therefore, there is a great need for energy trading paradigms to support FMENs for utilizing fully RESs in the energy production part. For this aim, this chapter suggests a transactive energy strategy for the optimal scheduling of full RESs microgrids in the grid.

There are several studies in the field of the optimal scheduling of microgrids in recent years. Each of them specialized in the related optimization problem by following diverse goals. The authors in [6] proposed a viable solution to employ a zero net energy community in the optimal scheduling of microgrids with customer-owned energy storage units. An energy management system strategy is proposed in [7] for the optimal scheduling of an institutional microgrid with the aim of increasing its self-consumption from clean decentralized units as well as reducing the operational costs considering the demand response programs. In [8], a new model with two planning and clustering layers is proposed for the day-ahead optimal scheduling of the microgrid in the presence of demand-side management, distributed generation resources, and energy storage systems. The authors presented a risk-constrained stochastic framework in [9] for joint reserve and energy scheduling of the microgrid considering demand-side management and its resiliency in both islanding and normal modes of the operation. In another work [10], the resilience-oriented scheduling of microgrids is conducted by introducing a robust-based model for the mixed-integer second-order cone programming considering the minimization of both load shedding and operational costs. Moreover, a novel microgrid resilient scheduling method is presented in [11] for preventing the potential frequency instability by including the system frequency dynamics and applying synthetic inertia control. In [12], the authors proposed a robust-based stakeholder-parallelizing distributed adaptive model for the scheduling of multi-stakeholders multi-microgrids to cope with ubiquitous uncertainties of diverse sources. Some other studies focus on multi-objective optimization frameworks for the optimal scheduling of microgrids. For example, in [13], a multi-objective optimization model is presented given the environmental and economic protection of microgrids with electric vehicles. In addition, the reference [14] proposed a practical multi-objective optimal scheduling approach to minimize the operation cost of microgrids and apply the chance-constrained programming technique for improving the user experience.

The evaluated studies indicate that different research works have been done for the optimal scheduling of microgrids under a variety of models.

However, none of them have looked to the scheduling of microgrids considering the grid modernization goals. In other words, developing a viable strategy and model that can provide confident conditions for the usage of 100 percent RESs in the microgrid is ignored in recent works. This issue has driven the main goal of the current chapter to offer transactive energy (TE) based strategy for microgrids in the presence of 100 percent RESs. Indeed, this work uses a TE for the optimal scheduling of 100 percent RESs microgrid aiming to balance energy in a reliable and sustainable manner. In order to increase the confidence of energy supply in the presence of 100 percent RESs, the hydrogen storage system (HS) is used to facilitate the energy conversion and storage processes for microgrids with wind power generation systems (WPGSs) and solar panels (SPs). Moreover, demand-side management programs are availed on the consumer-side to give an appropriate level of flexibility for microgrids in the scheduling procedure. Thus, the main goal of this chapter is to examine the capability of TE technology in easing the adoption of 100 percent RESs in hybrid energy networks. Due to this, the chapter is aimed to deeply investigate the proposed model in [15] for specifying how TE can be used in designing an energy sharing-based strategy for the 100 percent RESs-based system. Hence, in this chapter, the related problem formulation, simulation results, and discussions are scrutinized based on the model presented in [15].

9.2 Problem formulation

The objective function for the optimization problem in the optimal scheduling of microgrids is the minimization of total costs. This function includes the economic terms of operating different devices, energy sharing in the day-ahead market, and demand-side management programs. This objective function can be formulated as follows.

$$B_m = \sum_{s=1}^{S} \xi_s \cdot \left[\begin{array}{l} \sum_{i=1}^{I} \sum_{t=1}^{T} P_{i,t,s}^{Lsh} \cdot \lambda_t^{Lsh} + \sum_{t=1}^{T} \left[(c_m^{HP} + P_{t,m,s}^{HP} b_m^{HP} \right. \\ \qquad \left. + (P_{t,m,s}^{HP})^2 a_m^{HP}) + (b_m^{PH} + P_{t,m,s}^{PH} a_m^{PH}) \right] \\ + \sum_{t=1}^{T} \sum_{i=1}^{I} (P_{i,t,s}^{Lsh} - P_{i,t}^{Load}) \cdot \lambda_t^{Se} \cdot \Delta t - \sum_{t=1}^{T} P_{m,t}^{DA} \cdot \lambda_t^{DA} \cdot \Delta t \end{array} \right] \quad \forall m \tag{9.1}$$

where, λ_t^{Lsh}, λ_t^{Se}, and λ_t^{DA} are the prices of load shedding, energy selling to the consumer, and day-ahead energy sharing at time t. $P_{i,t,s}^{Lsh}$, $P_{t,m,s}^{HP}$, and $P_{t,m,s}^{PH}$ are the amounts of load shedding, hydrogen energy converted to the power, and the power converted to the hydrogen energy at time t, microgrid m, and scenario s. c_m^{HP}, b_m^{HP}, and a_m^{HP} are the cost coefficients for

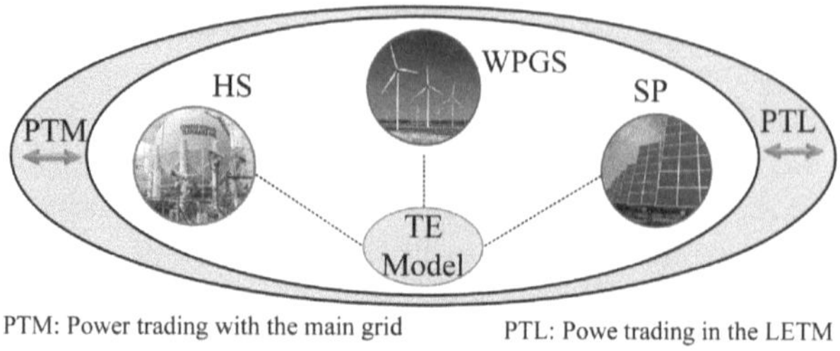

FIGURE 9.1 Schematic of the microgrid with RESs and storage devices.

the hydrogen to power process while b_m^{PH} and a_m^{PH} are the cost coefficients for the power to hydrogen energy conversion process in the HS. $P_{i,t}^{Load}$ and $P_{m,t}^{DA}$ are the power demand and day-ahead energy sharing. ξ_s and B_m are the probability of scenarios and the objective function.

The stochastic modeling of the microgrids' scheduling is done using the autoregressive integrated moving average technique (ARIMA) for scenario creation. Due to the complexity challenge of numerous scenarios, the fast forward selection method (FF) is availed for reducing the number of scenarios. Further information regarding the mathematical modeling of the ARIMA and FF methods can be found in [16,17] and [18,19].

Each microgrid in this study uses SPs and WPGSs to environmentally friendly meet energy demand and HS for benefiting energy conversion and storage possibility to increase reliability in the energy supply. The TE-based model is intended to make energy sharing among microgrids in the local energy trading market (LETM) possible and increase the flexibility of the system. Indeed, microgrids can deliver or receive energy from the LETM for balancing energy during the day. On the other hand, they can have energy sharing with the main grid for economic and technical objectives. The schematic of the microgrid with the related devices is indicated in Fig. 9.1.

In the optimal scheduling of microgrids, one of the main goals is to establish a dynamic power balance at all times. For satisfying this, the following energy balance constraint is considered in the optimization problem.

$$\left(P_t^{Loss} - \sum_{i=1}^{I}(P_{i,t}^{Lsh} - P_{i,t}^{Load})\right) + \sum_{m=1}^{M}(P_{m,t}^{ML} + P_{m,t}^{DA} + P_{m,t}^{PH}) = \sum_{m=1}^{M}(P_{m,t}^{LM} + P_{m,t}^{HP} + P_{m,t}^{SP} + P_{m,t}^{WPGS})\ \forall t \tag{9.2}$$

where, P_t^{Loss} is the losses related to the electrical energy. $P_{m,t}^{ML}$ is the power transmitted from the microgrid to the LETM and $P_{m,t}^{LM}$ is the power received by the microgrid from the LETM. $P_{m,t}^{SP}$ and $P_{m,t}^{WPGS}$ are solar and wind types of RESs for clean power production. In this work, RESs are mathematically modeled as follows.

$$P_{m,t}^{WPGS} = \begin{cases} 00 \le \varpi_t \le \varpi_m^{Cut-I} \\ P_m^{R,WP} \cdot (\gamma_1 \varpi_t^2 + \gamma_2 \varpi_t + \gamma_3) \varpi_m^{Cut-I} \le \varpi_t \le \varpi_m^{Rated} \\ P_m^{R,WP} \varpi_m^{Rated} \le \varpi_t \le \varpi_m^{Cut-O} \\ 0 \varpi_m^{Cut-O} \le \varpi_t \end{cases} \tag{9.3}$$

$$P_{m,t}^{WPGS} / \sqrt{(P_{m,t}^{WPGS})^2 + (Q_{m,t}^{WPGS})^2} = \tag{9.4}$$

$$R_t^{Solar} \cdot \eta_m^{SP} . Size_m^{SP} \ge P_{m,t}^{SP} \forall t, \forall m \tag{9.5}$$

where, $P_m^{R,WP}$ is the rated wind power and $\varpi_t$is its wind speed. γ_1,γ_2,and $\gamma_3$are the coefficients for the WPGS. The cut in, cut out, and rated speeds for the wind are respectively denoted by ϖ_m^{Cut-I}, ϖ_m^{Cut-O}, and ϖ_m^{Rated} . $Q_{m,t}^{WPGS}$ is the reactive wind power and is the constant. η_m^{SP} and $Size_m^{SP}$ are the efficiency and capacity of SPs. R_t^{Solar} is the solar radiation. Eqs. (9.3) and (9.4) model the WPGS [20] and Eq. (9.5) is for SP modeling [21].

As the necessary devices for accompanying RESs, energy storage systems are emerged to introduce a promising way for the energy balance of the system with stochastic producers. In this regard, hydrogen is one of the carriers of energy that offers suitable opportunities for energy conversion and storage when the system faces excess energy generation. Due to this, the HS is used for storing surplus generated energy and using it in other hours. The power to hydrogen process converts the power to the hydrogen energy for storing. Moreover, the hydrogen is converted back to power using a hydrogen-based gas turbine for feeding the power grid. The HS is operated based on the following formula [22].

$$H_{m,t}^{HES} - H_{m,t-1}^{HES} = P_{m,t}^{PH} \cdot \eta^{PH} - \frac{P_{m,t}^{HP}}{\eta^{HP}} \forall t, \forall m \tag{9.6}$$

$$\bar{P}_m^{PH} \cdot U_{m,t}^{PH} \ge P_{m,t}^{PH} \ge P_m^{PH} \cdot U_{m,t}^{PH} \tag{9.7}$$

$$\bar{P}_m^{HP} \cdot U_{m,t}^{HP} \ge P_{m,t}^{HP} \ge P_m^{HP} \cdot U_{m,t}^{HP} \tag{9.8}$$

$$1 \ge U_{m,t}^{HP} + U_{m,t}^{PH} \tag{9.9}$$

$$\bar{H}_m^{HES} \ge H_{m,t}^{HES} \ge H_m^{HES} \tag{9.10}$$

$$H_{m,T}^{HES} \le H_{m,End}^{HES}; H_{m,Initial}^{HES} = H_{m,0}^{HES} \tag{9.11}$$

where, $H_{m,t}^{HES}$ is the stored hydrogen in the HS. η^{HP} and η^{PH}($U_{m,t}^{HP}$ and $U_{m,t}^{PH}$) are the efficiency (binary variables) of hydrogen to power and vice versa process. Eq. (9.6) is for the hydrogen balance in the HS Eqs. (9.7) and (9.8) are the upper and lower bounds for the power to hydrogen and vice versa process. Eq. (9.9) is for the states of energy conversion in the HS Eqs. (9.10) and (9.11) are the limitations for the stored hydrogen in the HS

The demand-side management is one of the flexibility creation ways for microgrids that requires consumers' involvement in the process [23]. Here, both reaction of consumers to the price signals and load changes are considered as the demand response programs. The load shifting [24] and shedding programs [25] have the following formula in the mathematical modeling.

$$P_{i,t}^{Load} + P_{i,t}^{Lsh} = Sh_{i,t}^{L} + P_{i,t}^{L,Fo} \forall t, \forall i \tag{9.12}$$

$$\bar{P}_{i,t}^{Lsh} \geq P_{i,t}^{Lsh} \geq P_{i,t}^{Lsh} \tag{9.13}$$

$$\bar{P}_{i,t}^{L,Fo} \geq P_{i,t}^{L,Fo} - P_{i,t}^{Lsh} \geq P_{i,t}^{L,Fo} \tag{9.14}$$

$$0 \leq Sh_{i,t}^{L} \leq P_{i,t}^{L,Fo} \psi^{\max} \tag{9.15}$$

$$\sum_{t=1}^{T} Sh_{i,t}^{L} = 0 \tag{9.16}$$

where, $Sh_{i,t}^{L}$ and $P_{i,t}^{L,Fo}$ are the amounts of shifted and forecasted load. $\psi^{max}$is the coefficient for the maximum range of load shifting. Eq. (9.12) is for balancing energy load. Eqs. (9.13) and (9.14) are for modeling the load shedding program while the load shifting scheme is implemented by applying (9.15) and (9.16).

In addition to the demand-side management for flexibility, energy trading strategies are also recognized as one of the promising ways of providing flexibility. Here, participation in the day-ahead market is considered for the microgrids energy sharing with the main grid. However, the energy sharing in the day-ahead market needs to be kept in the allowable range according to the following constraint.

$$\bar{P}_{t}^{DA} \geq P_{m,t}^{DA} \geq P_{t}^{DA} \forall t, \forall m \tag{9.17}$$

This work offers a new way of energy sharing for microgrids by using the TE for creating the LETM aiming to allow microgrids for balancing energy under different energy transactions. This local area for energy sharing is designed based on the TE that enables the system to effectively use the energy sharing opportunity for sustainability. The following limitations are for implementing the LETM for transactive energy sharing.

$$1 \geq U_{m,t}^{LM} + U_{m,t}^{ML} \forall m, \forall t \tag{9.18}$$

$$U_{m,t}^{ML} \cdot \gamma^{ML} \geq P_{m,t}^{ML} \forall m, \forall t \tag{9.19}$$

$$U_{m,t}^{LM} \cdot \gamma^{LM} \geq P_{m,t}^{LM} \forall m, \forall t \tag{9.20}$$

$$\sum_{m} P_{m,t}^{ML} = \sum_{m} P_{m,t}^{LM} \forall t \tag{9.21}$$

$$\sum_{t} P_{m,t}^{ML} = \sum_{t} P_{m,t}^{LM} \forall m \tag{9.22}$$

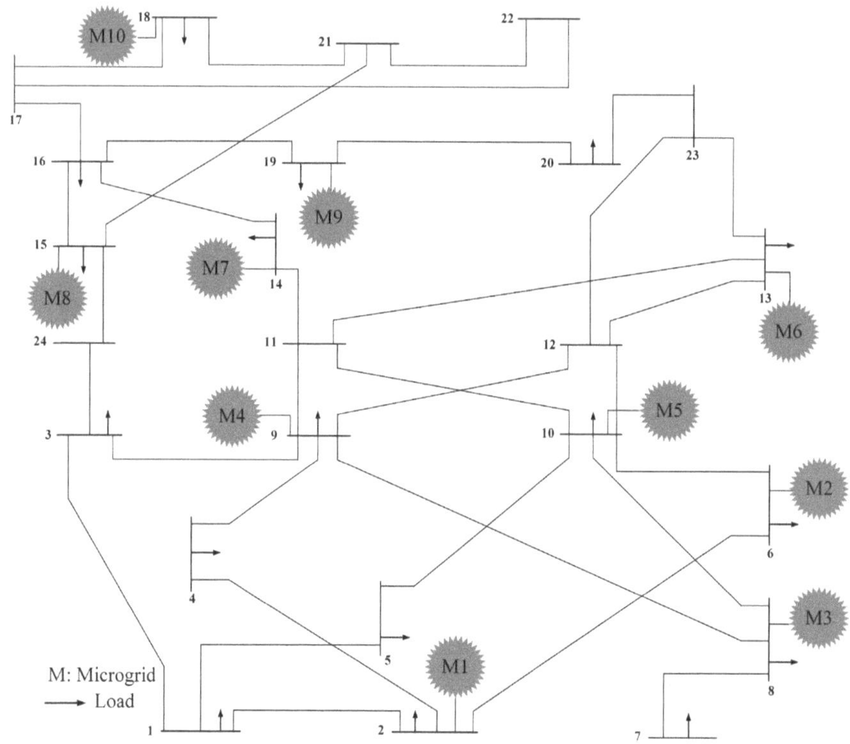

FIGURE 9.2 The schematic of the IEEE 24-bus test system.

where, $U_{m,t}^{LM}$ and $U_{m,t}^{ML}$ ($\gamma^{LM}$and γ^{ML}) are the binary variables (coefficients) for the states of energy sharing in the LETM Eq. (9.18) is for the energy transaction states in the LETM Eqs. (9.19) and (9.20) ((9.21) and (9.22)) are the allowable range of energy sharing (energy balance constraints) in the LETM

The AC power flow formulation is also intended for the electricity network modeling that the related mathematical models and descriptions can be found in [26].

9.3 Simulation results

This research renders a TE-based energy sharing model for scheduling microgrids in the hybrid grid with 100 percent RESs. Indeed, the possibility of full usage of RESs potential in power production is assessed in the networked microgrids. The IEEE 24-bus test system [27] with ten microgrids is selected for verifying the model's appropriateness. Fig. 9.2 denotes the schematic of this test system.

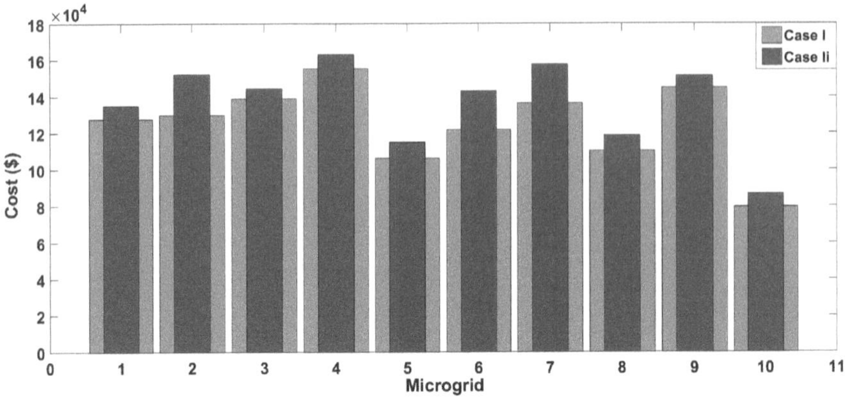

FIGURE 9.3 The total energy cost for microgrids in the optimal scheduling problem.

For analyzing the microgrids' optimal scheduling, ten microgrids with different devices are intended that their load data and energy price are available in [28]. WPGSs [20] and SPs [29] are two types of RESs that are used for power production. Moreover, microgrids use the HS for storing energy by converting power to hydrogen and vice versa [22]. The demand-side management programs are taken into account for making the microgrids more flexible. The energy sharing opportunities are also considered for microgrids, in which they can have purchasing or selling energy with the main grid. In addition to this, a TE-based strategy is offered for microgrids to share energy with each other in the LETM aiming to effectively handle the presence of 100 percent RESs.

Fig. 9.3 portrays the total energy cost for microgrids in the optimal scheduling problem. This work assesses the problem-solving in two cases, in which the only difference between them is related to the uncertainty modeling that is considered in Case II unlike Case I. According to Fig. 9.3, Case II presents more costs for microgrids in their operation under various uncertainties. In uncertainty modeling, considering the different stochastic changes can assist the modeling process to be near-real conditions. As both positive and negative changes are intended together, it results in more costs than deterministic modeling of the optimal scheduling of microgrids in Case I.

All microgrids consist of different energy production and storage systems as the basic equipment for the energy meeting. Here, Fig. 9.4 illustrates the output of the WPGS as well as the stored and generated energy in the HS. According to this figure, in the initial hours of the day, the HS is in the energy-storing state because of greater power production than consumption. However, it is operated to generate the power to feed the electric power system at peak times with a higher energy consumption rate.

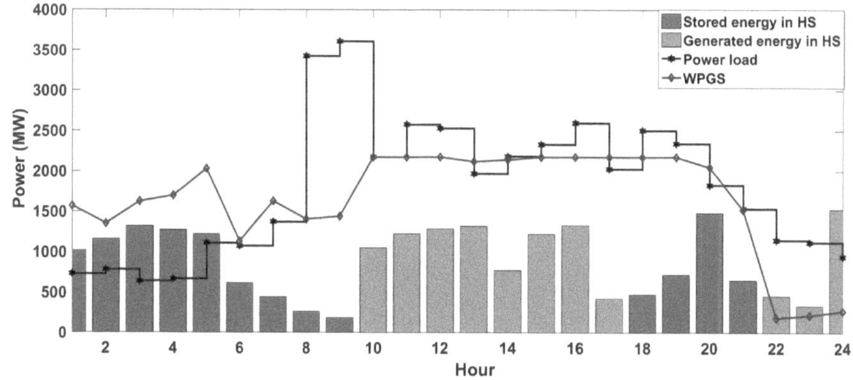

FIGURE 9.4 The output of the WPGS as well as the stored and generated energy in the HS.

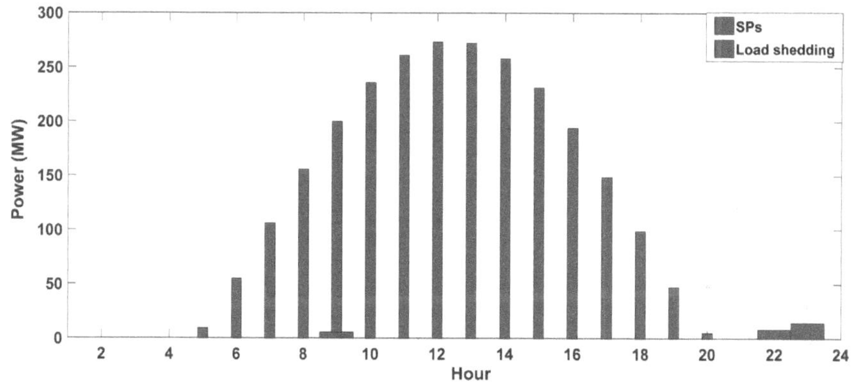

FIGURE 9.5 The output of SPs and the amount of load shedding in the microgrid.

In the evening (6 pm to 9 pm), reducing energy consumption has again led to charging the HS system. This is while the substantial reduction in the WPGS output in the last three hours of the day has driven the scheduling to use the HS potential for meeting a portion of power demand.

The SP is another RESs that uses solar energy for power production. In this study, SPs are deployed to cover the system's energy generation at peak hours when the grid faces a higher power consumption in comparison with other hours. Fig. 9.5 indicates the output of SPs and the amount of load shedding in the microgrid. This figure highlights that SPs have the maximum output in the mid hours of the day when microgrids require more energy for meeting their energy load. Thus, the effective potential of SPs is used for clean power production at peak times. On the other hand, since microgrids use fully RESs for power production, the system faces

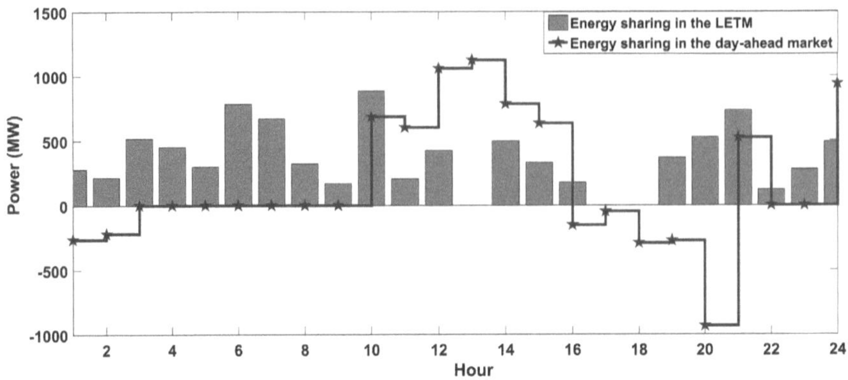

FIGURE 9.6 The amounts of microgrids' energy sharing with the power grid and in the LETM under the TE technology.

difficulties in having sufficient flexibility for reliable energy supply. Hence, demand-side management schemes are applied for improving flexibility. In this respect, microgrids use the load shedding program in emergency conditions for keeping the stability of the grid. Given Fig. 9.5, only in a few hours of the day, especially in the hours with the lowest WPGS output, microgrids used the load shedding possibility for balancing energy.

As already mentioned, flexibility is a critical factor for microgrids in the presence of 100 percent RESs to effectively be able for keeping the grid's stability. In addition to the demand-side management, energy sharing is considered as another suitable way for creating flexibility for microgrids. In this study, energy sharing with the main grid and in the LETM are intended for renewable-based microgrids. Fig. 9.6 demonstrates the amount of microgrids' energy sharing with the power grid and in the LETM under the TE technology. This figure states that microgrids have more purchased energy in the early morning and evening due to the lower price of energy in the mentioned hours. However, they sold a large amount of energy to the main grid to minimize their total energy costs at peak times with the higher energy price. Indeed, the limited energy sharing with the main grid allows microgrids to flexibly act in their optimal scheduling considering the technical and economic goals. On the other hand, microgrids used the energy sharing opportunity in the LETM under the TE mechanism. As seen in Fig. 9.6, almost in all times of the day, microgrids used the energy sharing in the local area to effectively benefit from the LETM potential in meeting energy demand. This way of interaction also improves the flexibility of microgrids in the presence of RESs for following their technical and economic objectives. Indeed, the TE technology enabled microgrids to have energy interactions in the form of energy sharing for keeping the stability of the grid by dynamically balancing energy.

9.4 Summary

The prominent goal of reaching 100 percent RESs grids has driven future modern energy networks to plan for practical strategies making the hybrid energy structure susceptible to welcome fully clean energy production. For this aim, this chapter investigated the key role of TE technology in supporting the energy grid to use 100 percent RESs benefits. The LETM was developed based on the TE technology to give the energy sharing opportunity for microgrids to meet energy demand and create a dynamic energy balance between the energy supply and demand. The microgrids were equipped with 100 percent RESs that bring a high level of uncertainties in energy production. Thus, this chapter considered the modeling of uncertainties by using the ARIMA and FF methods for scenario generation and reduction. The verification of the offered TE-based energy sharing strategy was done by selecting the modified IEEE 24-bus test system as the case study. The problem was solved in two cases with uncertainties effects and a deterministic state. Given the obtained results, it is concluded that the TE technology can be used as a promising way of enabling the energy grid to adopt 100 percent RESs in an affordable and reliable manner. Indeed, it can improve the system's conditions for using fully RESs while supporting it for keeping the sustainability of the energy structure.

9.5 Questions

The careful analyzing whole of this chapter is expected to enable readers to effectively evaluate the following questions.

1) What are the key motivations for transiting toward the 100 percent RESs in future modern energy networks?
2) What are the benefits and challenges of the energy structure with 100 percent RESs?
3) Why energy conversion and storage technologies are necessary for the realization of 100 percent RESs plans?
4) From which ways the sufficient flexibility can be provided for the energy grid with 100 percent RESs?
5) What is the vital role of TE technology in facilitating the adoption of fully RESs in the energy generation sector?
6) How can TE establish affordable and reliable energy sharing opportunities for market members in the hybrid system?
7) How can demand-side management schemes assist for the 100 percent presence of RESs in the energy landscape?
8) How a high level of uncertainties can be managed and stochastically modeled in the renewable-based energy grid?

9) What are the necessities for designing the LETM in creating the local energy sharing possibility for the market players?
10) What are the benefits of energy sharing at the local level?

9.6 Suggestions

Given the topic of this chapter, the following suggestions are provided for the readers assisting them to pursue related research clues.

1) Developing TE-based energy sharing strategies can provide appropriate conditions for using 100 percent RESs in the energy production sector of future energy networks.
2) Upgrading energy conversion and storage technologies can facilitate the adoption of fully RESs for clean energy production in the hybrid energy network.
3) Designing innovative demand-side management programs can support energy grids with 100 percent RESs in keeping its reliability and sustainability in energy supply.

9.7 Future trends

Future trends can give a proper overview for the readers to form their related works and develop effective solutions for the remaining gaps, which are listed as follows for this chapter.

1) Novel energy sharing strategies need to be developed for making the plans for 100 percent RESs implementable in the practice.
2) Energy conversion and storage systems are the necessary components of the energy grid that should be incorporated into the system with 100 percent RESs.
3) Demand-side management programs need to be developed in a way to improve the flexibility of the consumer-side for effectively dealing with the negative effects of uncertainties in the system.

References

[1] D.V. Pombo, J. Martinez-Rico, H.M. Marczinkowski, Towards 100% renewable islands in 2040 via generation expansion planning: the case of São Vicente, Cape Verde, Appl. Energy 315 (2022) 118869.
[2] A. Rahman, O. Farrok, M.M. Haque, Environmental impact of renewable energy source based electrical power plants: solar, wind, hydroelectric, biomass, geothermal, tidal, ocean, and osmotic, Renew. Sustain. Energy Rev. 161 (2022) 112279.
[3] M. Daneshvar, B. Mohammadi-Ivatloo, K. Zare, M. Abapour, S. Asadi, A. Anvari-Moghaddam, Chance-constrained scheduling of hybrid microgrids under transactive energy control, Int. J. Energy Res. 45 (7) (2021) 10173–10190.

[4] M. Daneshvar, M. Abapour, B. Mohammadi-ivatloo, S. Asadi, Impact of optimal DG placement and sizing on power reliability and voltage profile of radial distribution networks, Majlesi J. Electr. Engineer. 13 (2) (2019) 91–102.

[5] W. Wang, Y. Huang, M. Yang, C. Chen, Y. Zhang, X. Xu, Renewable energy sources planning considering approximate dynamic network reconfiguration and nonlinear correlations of uncertainties in distribution network, Int. J. Electr. Power Energy Syst. 139 (2022) 107791.

[6] H.-C. Jo, G. Byeon, J.-Y. Kim, S.-K. Kim, Optimal Scheduling for a Zero Net Energy Community Microgrid With Customer-Owned Energy Storage Systems, IEEE Trans. Power Syst. 36 (3) (2020) 2273–2280.

[7] H.A.U. Muqeet, A. Ahmad, Optimal scheduling for campus prosumer microgrid considering price based demand response, IEEE Access 8 (2020) 71378–71394.

[8] A. Bolurian, H. Akbari, S. Mousavi, Day-ahead optimal scheduling of microgrid with considering demand side management under uncertainty, Electr. Power Systems Res. 209 (2022) 107965.

[9] M. Vahedipour-Dahraie, H. Rashidizadeh-Kermani, A. Anvari-Moghaddam, Risk-based stochastic scheduling of resilient microgrids considering demand response programs, IEEE Syst. J. 15 (1) (2020) 971–980.

[10] N.-M. Zografou-Barredo, C. Patsios, I. Sarantakos, P. Davison, S.L. W.alker, P.C. T.aylor, MicroGrid resilience-oriented scheduling: a robust MISOCP model, IEEE Trans. Smart Grid 12 (3) (2020) 1867–1879.

[11] Z. Chu, N. Zhang, F. Teng, Frequency-Constrained Resilient Scheduling of Microgrid: a Distributionally Robust Approach, IEEE Trans. Smart Grid 12 (6) (2021) 4914–4925.

[12] H. Qiu, W. Gu, Y. Xu, Z. Wu, S. Zhou, G. Pan, Robustly multi-microgrid scheduling: stakeholder-parallelizing distributed optimization, IEEE Trans. Sustain. Energy 11 (2) (2019) 988–1001.

[13] Y. Mei, B. Li, H. Wang, X. Wang, M. Negnevitsky, Multi-objective optimal scheduling of microgrid with electric vehicles, Energy Rep. 8 (2022) 4512–4524.

[14] H. Chen, L. Gao, Z. Zhang, Multi-objective optimal scheduling of a microgrid with uncertainties of renewable power generation considering user satisfaction, Int. J. Electr. Power Energy Syst. 131 (2021) 107142.

[15] M. Daneshvar, B. Mohammadi-Ivatloo, K. Zare, S. Asadi, Transactive energy management for optimal scheduling of interconnected microgrids with hydrogen energy storage, Int. J. Hydrogen Energy 46 (30) (2021) 16267–16278.

[16] M. Daneshvar, B. Mohammadi-Ivatloo, K. Zare, A novel transactive energy trading model for modernizing energy hubs in the coupled heat and electricity network, J. Clean. Prod. 344 (2022) 131024.

[17] J. Contreras, R. Espinola, F.J. N.ogales, A.J. C.onejo, ARIMA models to predict next-day electricity prices, IEEE Trans. Power Syst. 18 (3) (2003) 1014–1020.

[18] M. Daneshvar, B. Mohammadi-ivatloo, S. Asadi, S. Galvani, Short term optimal hydro-thermal scheduling of the transmission system equipped with pumped storage in the competitive environment, Majlesi J. Electr. Engineer. 14 (1) (2020) 77–84.

[19] L. Wu, M. Shahidehpour, T. Li, Stochastic security-constrained unit commitment, IEEE Trans. Power Syst. 22 (2) (2007) 800–811.

[20] R. Jabbari-Sabet, S.-M. Moghaddas-Tafreshi, S.-S. Mirhoseini, Microgrid operation and management using probabilistic reconfiguration and unit commitment, Int. J. Electr. Power Energy Syst. 75 (2016) 328–336.

[21] M. Daneshvar, B. Mohammadi-ivatloo, K. Zare, Integration of distributed energy resources under the transactive energy structure in the future smart distribution networks, Operation of Distributed Energy Resources in Smart Distribution Networks, Elsevier, 2018, pp. 349–379.

[22] M. Ban, J. Yu, M. Shahidehpour, Y. Yao, Integration of power-to-hydrogen in day-ahead security-constrained unit commitment with high wind penetration, J. Mod. Power Syst. Clean Energy 5 (3) (2017) 337–349.
[23] M.K. K.iptoo, O.B. A.dewuyi, M.E. L.otfy, A.M. I.brahimi, T. Senjyu, Harnessing demand-side management benefit towards achieving a 100% renewable energy microgrid, Energy Rep. 6 (2020) 680–685.
[24] M. Daneshvar, H. Eskandari, A.B. S.irous, R. Esmaeilzadeh, A novel techno-economic risk-averse strategy for optimal scheduling of renewable-based industrial microgrid, Sustain. Cities Soc. 70 (2021) 102879.
[25] J. Qiu, J. Zhao, H. Yang, Z.Y. D.ong, Optimal scheduling for prosumers in coupled transactive power and gas systems, IEEE Trans. Power Syst. 33 (2) (2017) 1970–1980.
[26] M. Daneshvar, B. Mohammadi-Ivatloo, A. Anvari-Moghaddam, Optimal energy trading strategy for proactive DISCO considering demand response programs in the distribution networks, in: CIGRE Symposium Aalborg 2019, CIGRE (International Council on Large Electric Systems, 2019.
[27] M.R. S.alehizadeh, A. Rahimi-Kian, M. Oloomi-Buygi, Security-based multi-objective congestion management for emission reduction in power system, Int. J. Electr. Power Energy Syst. 65 (2015) 124–135.
[28] Y. Chen, M. Hu, Balancing collective and individual interests in transactive energy management of interconnected micro-grid clusters, Energy 109 (2016) 1075–1085.
[29] M. Daneshvar, B. Mohammadi-Ivatloo, S. Asadi, M. Abapour, A. Anvari-Moghaddam, A transactive energy management framework for regional network of microgrids, in: 2019 International Conference on Smart Energy Systems and Technologies (SEST), IEEE, 2019, pp. 1–6.

CHAPTER

10

An application of GAMS in simulating hybrid energy networks optimization problems

10.1 Introduction

The movement of modern society is formed to effectively assess the current challenges in human life aiming to draw specialized trends for addressing them by establishing harmonious functioning. However, first of all, we need to know what are the requirements for developing the aforementioned trends and harmonious functioning. In the energy sector, considering this question has led to the transition of focuses from traditional systems to intelligent devices. All of these advancements are acted as strong pushing leverage for progressing the information and communication technologies and different smart control systems, objects, appliances, etc. Herein, the prominent point that we have to pay attention to it is that all of the aforementioned progresses and developments were only a prototype before any changes and growth. Indeed, for any development, we need to completely identify its requirements and provide them accordingly. The energy sector is no exception [1]. In the energy network, after the idea about all parts of the system is formed, it needs to be mathematically developed and simulated for analyzing the results after implementing it in practical cases [2]. We have fully discussed the conceptual parts of future modern multi-carrier energy networks (MCENs) and transactive energy technology in the first seven chapters, presented the mathematical

Emerging Transactive Energy Technology for Future Modern Energy Networks

DOI: https://doi.org/10.1016/B978-0-323-91133-7.00009-0

models for the most common parts of the modern MCENs in Chapters 6 and 7, and now it's time to focus on simulation of them. For effectively simulating the energy systems, capable software is necessary with strong tools and complete input data for achieving confident results [3]. On the other hand, hybrid energy networks with numerous systems and complex structures need to be optimized from the viewpoint of different aspects [4]. This is key for the modern MCENs to work based on the best possible set points that for the first time, they are going to be extracted from suitable software [5]. Therefore, a great need is felt for the full option optimization software with strong tools that can not only enable the system to effectively optimize the modern MCENs but also can appropriately solve the different types of mathematical problems for extracting optimal solutions. In this regard, the General Algebraic Modeling System (GAMS) is introduced as a high-level modeling system for optimization and mathematical programming [6]. Indeed, GAMS is specially developed for solving the different types of optimization problems without limitations in the size and type of the problem. It consists of a stable of integrated high-performance solvers along with a language compiler that allows the user to thoroughly build his/her model for simulation. In modern MCENs, there are several network-based problems with a large size in most cases that need to be effectively optimized for developing a sustainable system [7]. In such a structure, employing appropriate optimization software is necessary for solving large-scale problems with numerous parameters and variables. Therefore, GAMS as one of the powerful optimization software can be widely used for solving the different problems of modern MCENs. This is the main motivation for us to effectively describe how we can easily use the GAMS in our optimization problems. There are several effective resources for teaching the GAMS software that have fully taught GAMS materials along with some practical-based examples such as references [8]. Because reading a complete book may be very time-consuming for readers, we have prompted to briefly and clearly describe how we can learn to work with GAMS in the shortest time. Therefore, this chapter is developed to do this in the best manner.

10.2 What is GAMS?

GAMS is comprehensive optimization software that facilitates the turning of real phenomena into mathematical models. It is tailored for large-scale, complex, and mathematical modeling applications, which allows the user to develop large maintainable models that are rapidly adaptable to new situations [9]. GAMS is a strong modeling tool that is specially designed for modeling different types of optimization problems such as linear programming (LP), nonlinear programming (NLP) [10], and mixed-integer linear programming (MILP) [4], and mixed-integer nonlinear programming (MINLP) [11].

10.3 Features of GAMS

GAMS is designed in a way to completely rely on its core competence and is mainly focused on empowering the users for building the most readable and maintainable models by developing the best solvers available anywhere [9]. GAMS is famous for its significant features that some of the most important of them are as follows.

- Independence of model and data [9]: GAMS allows for writing the model independently of the data. It supports the different kinds of data sources such as Access, Excel, or GDX (GAMS Data eXchange) file format. A GDX file has the capability of storing various GAMS symbols such as Eqs., variables, parameters, or sets. Indeed, GDX files have made the operation of data for a GAMS model, presentation of a GAMS model results, and storing of the same model results possible in the best manner. However, it is not structured to store executable statements or model formulation.
- Independence of model and platform [9]: GAMS enables the models to be fully portable among platforms – write once, run anywhere. GAMS can be run on IBM Power AIX, Sparc Solaris, SOLARIS, Mac OS X, Linux, and Windows.
- Independence of model and user interface [9]: The GAMS object-oriented APIs enable the GAMS for seamless integration into an application by providing proper classes for engaging with GAMS. The object-oriented GAMS API is available in three versions: Python, Java, and NET work with Python 3.4, 2.7 and 2.6, Java SE 5 and up, and .NET framework 4 (Visual Studio 2010). Moreover, GAMS provides smart links for different applications such as MatLab, MS Excel, or R.
- GAMS is structured based on the common English words that enable the users to easily understand various statements and work with it.
- Given the independence of the model with the solution method, GAMS offers the possibility of solving one model using diverse solution methods only by changing the solver.
- Implementing the mathematical Eqs. in GAMS is transparent due to building a simulation environment that resembles mathematical programming models.
- Capability of solving the small and large-scale problems with small code by effectively using the index applications.

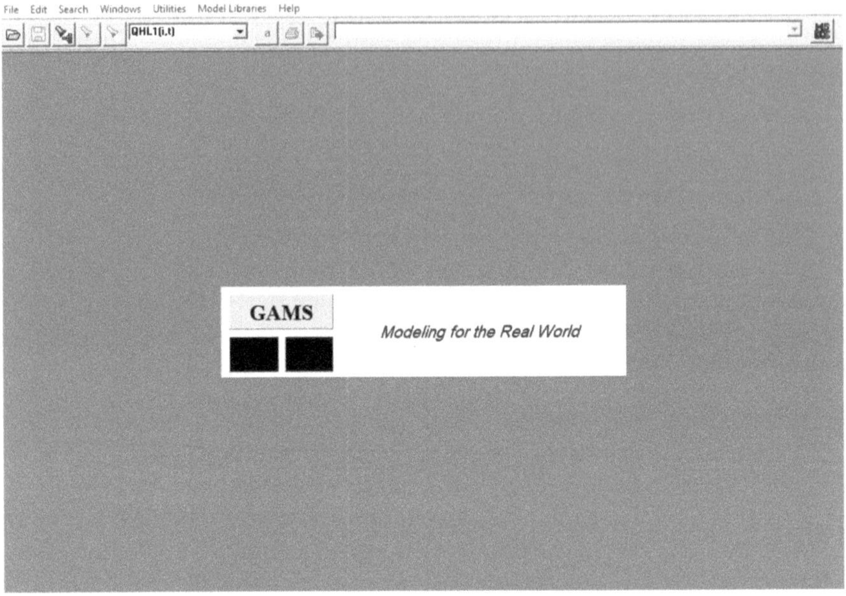

FIGURE 10.1 The schematic of the GAMS work environment.

10.4 How can I install GAMS on my PC?

The free version of GAMS can be downloaded from the following link [12]:

GAMS download link: https://www.gams.com/download/

In a demo license, model size is limited as follows [12]:

- For linear models, GAMS will generate and solve models with up to 2000 variables and 2000 constraints.
- For all other model types, GAMS will generate and solve models with up to 1000 variables and 1000 constraints.

After downloading and installing GAMS on a PC, when we open GAMS, we will face the GAMS work environment illustrated in Fig. 10.1.

The user can create a new project in the desired place on his/her PC by going to the *File* menu and selecting *New Project*. After creating a new project, several GAMS files can be created for a certain project by going to the *File* menu and selecting the *New* option. When we select the New option, the GAMS file is created that is the environment we can write our code and implement our models in it. The users are recommended to *Save* each change after they are applied in the model and *Save as* its related file in a suitable place. The Save and Save as options can be accessed in the tab *File*.

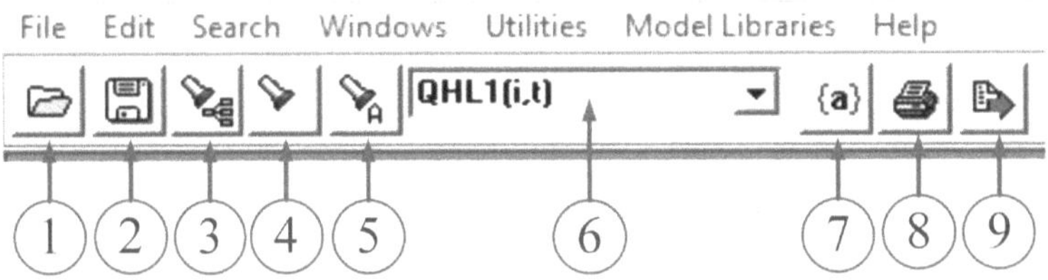

FIGURE 10.2 The schematic of the GAMS work environment.

10.5 Some important points before working with GAMS

Once we are familiar with how to create a project and the GAMS file within the project, we need to go through the process of creating a model in the GAMS environment. However, before starting this process, it is better to review several important points that are given as:

- We need to thoroughly understand the mathematical model before implementing it in GAMS. This can be helpful for reducing errors during model development and extracting logical results.
- The Enter key is used for going to the next line.
- The "*" is used for adding descriptions in the model that has not any impact on the solving process and results. Also, we can use the "$ontext" command for starting to add descriptions and the "$offtext" command for ending descriptions in the model.
- There is no difference between uppercase and lowercase letters in GAMS.
- Each GAMS program consists of several sections. Several parts in each model are separated from each other using ";" in GAMS.
- In GAMS, the GAMSIDE makes the accessibility to the documentation possible. In this respect, we can access the different GAMS online and offline documentations by referring to the Help tab.

In the toolbar of GAMS, there are several applicable keys that are illustrated in Fig. 10.2.

The descriptions of the different keys in the GAMS toolbar is as follows.

1. This key can be used for opening the stored GAMS code on the PC.
2. This key can be used for storing the in-hand code in GAMS.
3. This key can be used for finding the text in the different files.
4. This key can be used for finding the text inside the code.
5. This key can be used for refinding the text inside the code.
6. This part allows the user to write the related text in the box for searching inside the code.

TABLE 10.1 Several common characters in GAMS [13].

A to Z	Alphabet	()	Parenthesis
&	Ampersand	[]	Square brackets
*	Multiply	) (	Braces
@	at	%	Percent
\	Back slash	0 to 9	Numerals
:	Colon	#	Pound sign
,	Comma	?	Question mark
$	Dollar	;	Semicolon
.	Dot	'	Single quote
+	Plus	/	Slash
a to z	Alphabet	→ ←	Space
"	Double quote	_	Underscore
=	Equal	!	Exclamation mark
>	Greater than	^	Circumflex
<	Less than	–	Minus

7. Inside our models, we may use numerous parentheses that are usually led to the parentheses error. Due to avoid this error, we can easily use this key to identify are the opened parenthesis closed or not.
8. This key can be used for printing the written code.
9. After completing the mode, this key allows us to run the program.

10.6 Terms, symbols, and reserved words in GAMS

In this section, some commonly used terms, symbols, and reserved words are provided [13]. The common characters used in GAMS are listed in Table 10.1.

Similar to other programming languages such as Java and C, GAMS uses reserved words (often also called keywords) that are already predefined. Some of these reserved words are listed in Table 10.2 [13].

Moreover, the most common operators in GAMS are tabulated in Table 10.3 [13].

10.7 Common mathematical functions in GAMS

In GAMS, mathematical functions enable the user to easily implement his/her model. The most common mathematical functions in GAMS are described here [13].

TABLE 10.2 Some predefined reserved words in GAMS [13].

Abort	**Acronym**	**Lt**	**Maximizing**
assign	binary	Ne	
Eq.	Eqs.	Ord	Parameter
No	option	Scalars	Set
Using	variable	Sos2	Sum
Prod	Scalar	Variables	Or
Smin	Sos1	If	Then
Until	While	Files	Putpage
semiint	File	Solve	for
models	Na	all	And
options	Or	eps	Eq
Le	Loop	Inf	Integer
Acronyms	Alias	Minimizing	Model
Card	Display	Negative	Not
Ge	Gt	Sets	Positive
System	Smax	Else	Repeat
Yes	Table	put	Semicont

TABLE 10.3 Most common operators in GAMS [13].

Operator	**Example**	
**	$a^{**}b(a > 0)$	a^b
*	a^*b	$a \times b$
/	a/b	$\frac{a}{b}$
-, +	$a \pm b$	$a \pm b$

10.7.1 Abs

This function can be used for calculating the absolute value of a term or expression. By using this function in Eqs., the model type will be NLP. Fig. 10.3 shows the example for the Abs function in GAMS.

10.7.2 Exp

This function can be used for calculating the exponentiation value of a term or expression. By using this function in Eqs., the model type will be NLP. Fig. 10.4 illustrates the example of the Exp function in GAMS.

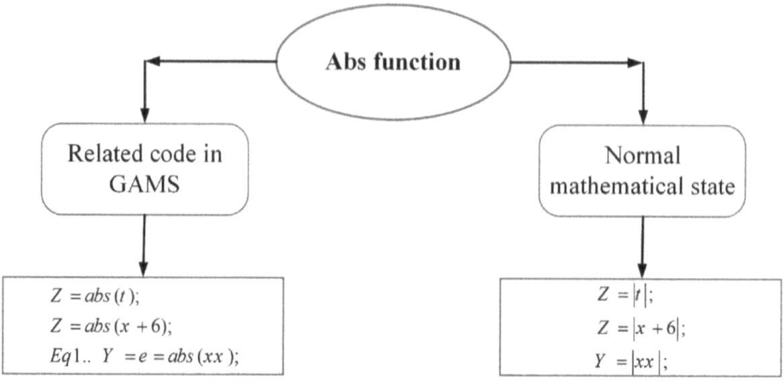

FIGURE 10.3 The example for the Abs function in GAMS.

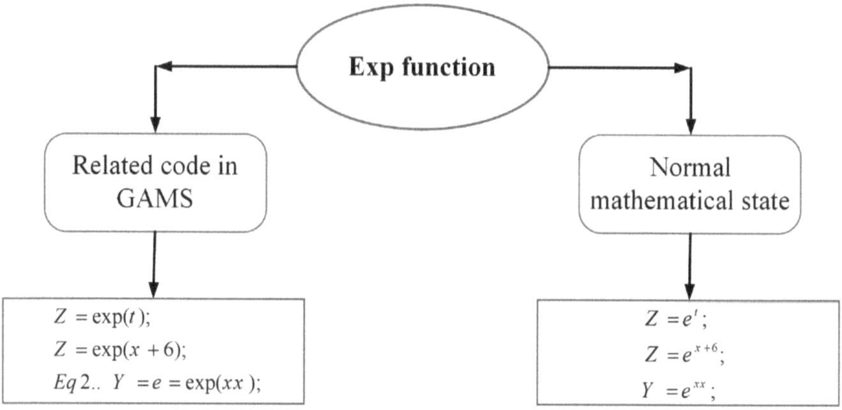

FIGURE 10.4 The example for the Exp function in GAMS.

10.7.3 Log, log10

These functions can be used for calculating the natural logarithm and logarithm base 10 value of a term or expression. By using these functions in Eqs., the model type will be NLP. Fig. 10.5 demonstrates the example of the Log and Log10 functions in GAMS.

10.7.4 Max, Min

These functions can be used for calculating the maximum and minimum value of a set of terms or expressions. By using these functions in Eqs., the model type will be the discontinuous nonlinear program (DNLP) [14]. Fig. 10.6 depicts the example of the Max and Min functions in GAMS.

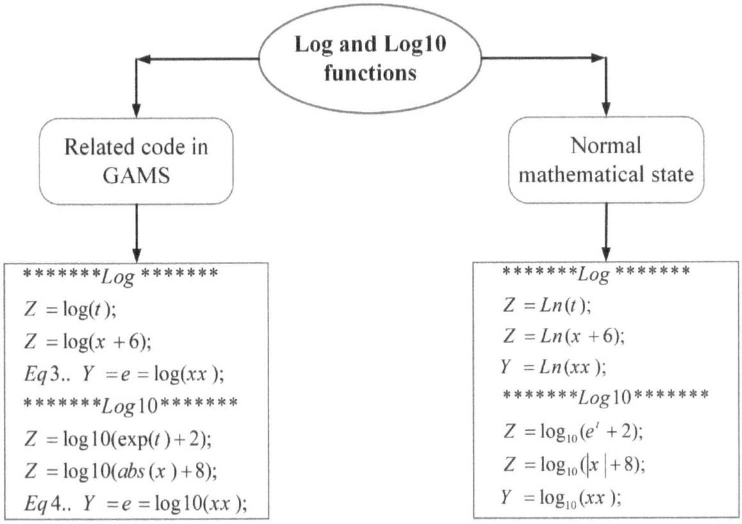

FIGURE 10.5 The example for the Log and Log10 functions in GAMS.

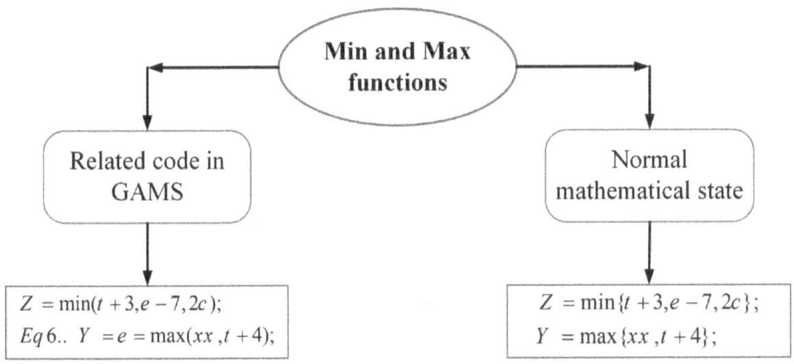

FIGURE 10.6 The example for the Max and Min functions in GAMS.

10.7.5 Prod

This function can be used for calculating the product of set indexed terms or expressions. By using this function in Eqs., the model type will be NLP. Fig. 10.7 shows the example of the Prod function in GAMS.

10.7.6 Round

This function can be used for rounding the numerical result of a term or expression. This function can be used in the data part of GAMS and it cannot be typically used in models. The Round function exists in two variants. The first variant of it rounds the numerical result to the nearest integer value while the second one rounds the numerical result to the

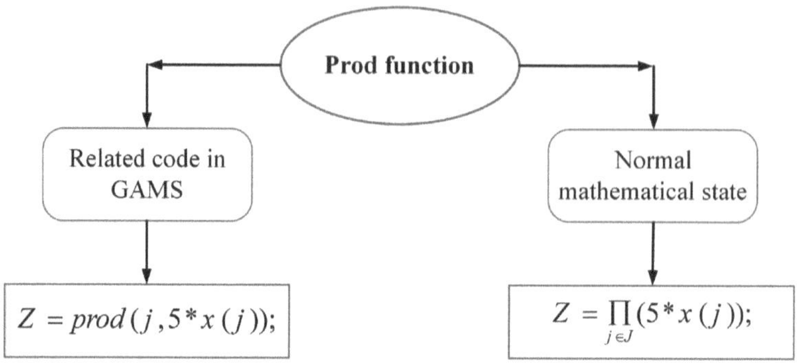

FIGURE 10.7 The example for the Prod function in GAMS.

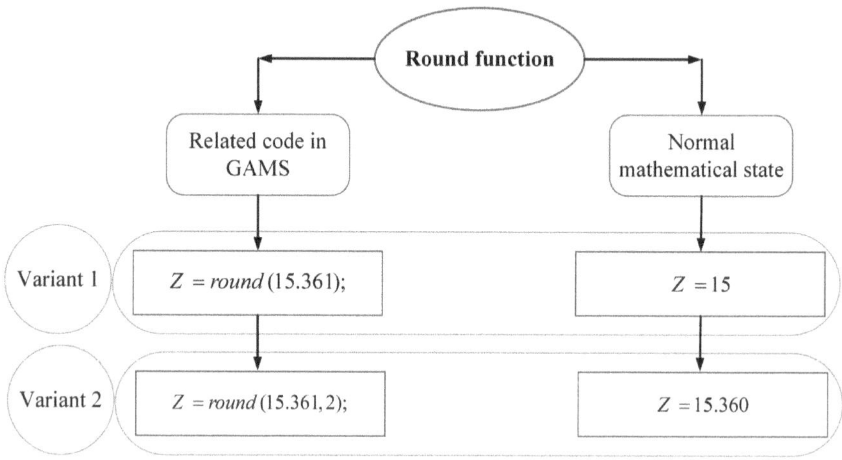

FIGURE 10.8 The example for the Round function in GAMS.

number of decimal points specified by the second argument. Fig. 10.8 illustrates the example of two variants of the Prod function in GAMS.

10.7.7 Smax, Smin

These functions can be used for calculating the maximum and minimum of set indexed terms or expressions. By using these functions in Eqs., the model type will be DNLP. Fig. 10.9 demonstrates the example of Smax and Smin functions in GAMS.

10.7.8 Sqr

This function can be used for calculating the square of a term or expression. By using this function in Eqs., the model type will be NLP. Fig. 10.10 depicts the example of the Sqr function in GAMS.

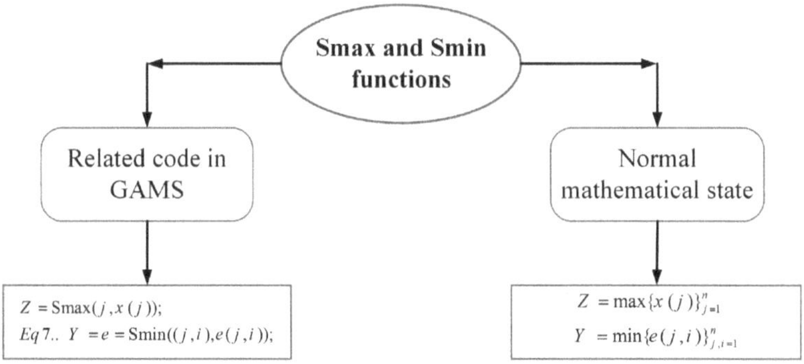

FIGURE 10.9 The example for the Smax and Smin functions in GAMS.

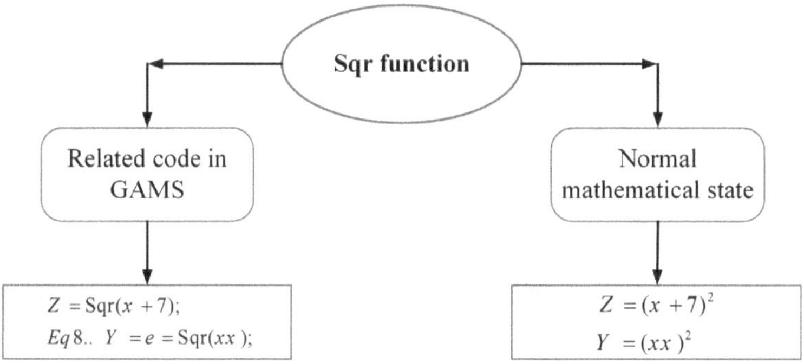

FIGURE 10.10 The example for the Sqr function in GAMS.

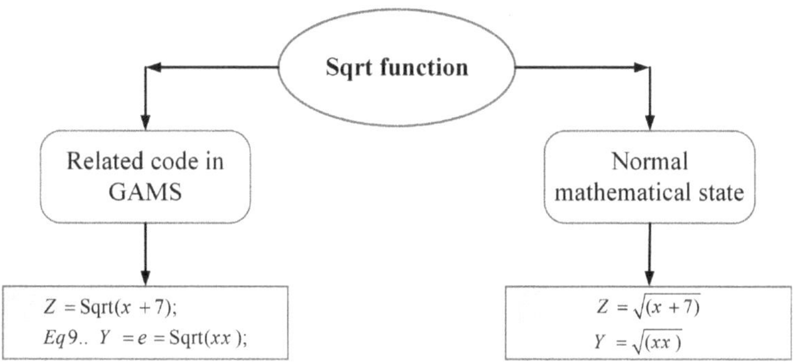

FIGURE 10.11 The example for the Sqrt function in GAMS.

10.7.9 Sqrt

This function can be used for calculating the square root of a term or expression. By using this function in Eqs., the model type will be NLP. Fig. 10.11 shows the example of the Sqrt function in GAMS.

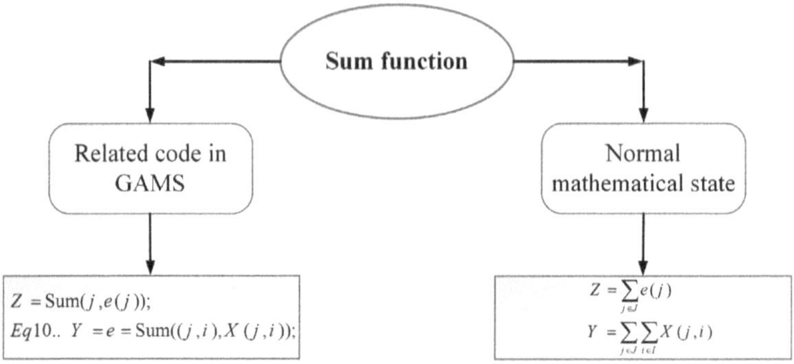

FIGURE 10.12 The example for the Sum function in GAMS.

10.7.10 Sum

This function can be used for calculating the sum of set indexed terms or expressions. By using this function in Eqs., the model type will be usually LP. It can be used on parameters or variables in models or on data. Fig. 10.12 illustrates the example of the Sum function in GAMS.

10.7.11 Other mathematical functions in GAMS

In GAMS, there are other useful mathematical functions that make the implementation of the different mathematical models easier. Some most important of them are listed in Table 10.4 [13].

10.8 GAMS commands

Generally, each GAMS program consists of nine main commands that we are going to describe each of these commands with an appropriate example in this section. In addition to these nine main commands, there are some sub-commands that can be used in GAMS models. In order to make GAMS learning more applicable, we have written the examples for each command in GAMS model considering the mathematical models provided in Chapters 6 and 7.

10.8.1 Set

In GAMS, SETS are typically used for defining indices that have names and are written through a combination of digits and letters. Defining SET is the first step in developing a certain model in GAMS. To this end, we can use the general format of SET for defining it based on Fig. 10.13.

For example, in Chapters 6 and 7, almost all mathematical models provided for most of the devices in the MCENs are considered for several

TABLE 10.4 Some of the common mathematical functions in GAMS [13].

Function	Description
Sin (x)	Sinose of the argument x where x should be in radians.
Cos (x)	Cosine of the argument x where x should be in radians.
ArcSin (x)	Arc sine of the argument x.
ArcCos (x)	Arc cosine of the argument x.
ArcTan (x)	Arc tangent of the argument x.
Sinh (x)	Hyperbolic sinose of the argument x.
Cosh (x)	Hyperbolic cosine of the argument x.
Edist (x,y,z,...)	Sum of squared values calculation= $x^2 + y^2 + z^2 + \ldots$
Fact (x)	Factorial of x where x is an integer.
Normal (x,z)	Random number normally distributed with mean x and standard deviation z.
Floor (x)	Largest integer that is less than or equal to x.
Pi	Value of Pi=3.141716...
Sign (x)	$\text{Sign}(x) = \begin{cases} 1x > 0 \\ 0x = 0 \\ -1x < 0 \end{cases}$
Power (x,z)	x raised to an integer power x^y,where z should be an integer.
Uniform (x,z)	Random number with a uniform distribution between x and z.
Uniformint (x,z)	Integer random number with a uniform distribution between x and z.

Set

Set name Optional defining text /Defining the elements related to the Set/

FIGURE 10.13 The definition of the Set command in GAMS.

interconnected microgrids and their operating is intended for time t. If we want to model such mathematical formulations in GAMS, we need to define the number of microgrids and time using Set for GAMS. For instance, assume that we want to optimally schedule the renewable-based system with ten microgrids for 24-hours. For such a problem, the microgrid, time, and bus indices can be defined in the GAMS environment as illustrated in Fig. 10.14.

Given Fig. 10.14, two points can be described as:

1. *m2(i)* is a subset of the bus index i that has made the separating of the buses connected to the microgrids with other buses in the study easier.

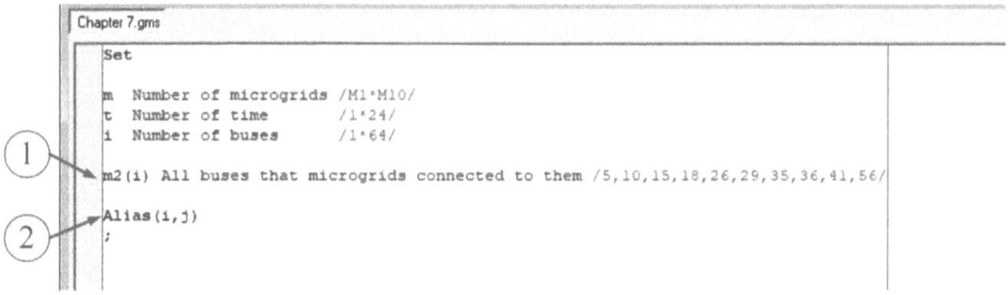

FIGURE 10.14 The example for the SET command in GAMS.

Scalar

Item name Optional labeling text /Numerical value/

FIGURE 10.15 The definition of the Scalar command in GAMS.

Depending on the type of the problem, GAMS allows for defining multi-dimensional Sets (sets with up to 20 dimensions).

2. On some occasions, we may need to address a single set more than once in a statement. For example, in the AC power flow formulation, we have *i* and *j* as two indices that refer to the network buses in the study. For such cases in GAMS, we can easily use the ALIAS command.

10.8.2 Data

In GAMS, arrays (scalar, zero, and parameters or multidimensional matrices) contain digital data. Defining SETs is necessary for the data part because they play an indices role for the aforementioned arrays. Declaring an array to contain data can be done in three formats in GAMS that are given as:

- Scalar (Zero-dimensional)
- Parameter (One-dimensional)
- Table (Multi-dimensional)

In this respect, Scalar, Parameter, and Table can respectively cover the scalars, vectors, and matrices.

10.8.2.1 Scalar

A Scalar is a type of Parameter that has not any dimension. Indeed, Scalar allows the user to define the scalars in the related problem. Let's back to our mathematical models in Chapters 6 and 7, for example, in the battery energy storage system (BESS) modeling, we need to define what is the efficiency of the BESS or PV panels and other similar scalars. In GAMS, we can easily do this by defining scalars. The general format of the Scalar command is illustrated in Fig. 10.15.

```
Set

m  Number of microgrids /M1*M10/
t  Number of time       /1*24/
i  Number of buses      /1*64/

m2(i) All buses that microgrids connected to them /5,10,15,18,26,29,35,36,41,56/

Alias(i,j)
;


Scalar

hBc   Charging efficiency of the BESS          /0.90/
hBd   Discharging efficiency of of the BESS    /0.90/
hPV   Efficiency of the PV panel               /0.22/
;
```

(1) → hBc; (2) → hPV

FIGURE 10.16 The example for the Scalar command in GAMS.

Parameter

Item name (set dependency) Optional labeling text / first settlement name associated value

second settlement name associated value

/;

FIGURE 10.17 The definition of the Parameter command in GAMS.

One simple example related to the mathematical models in Chapters 6 and 7 is demonstrated in Fig. 10.16.

According to Fig. 10.16, we have the following points:

1. Presents the efficiency of charging and discharging of the BESS.
2. Presents the amount of PV panels efficiency.

10.8.2.2 Parameter

Given the indices defined in the Set, Parameter format can be used for entering one-dimensional data in GAMS. All parameters have zero amounts in default and we can enter the intended amount to each parameter according to the Parameter general format shown in Fig. 10.17.

As an example, the amount of solar radiation in each hour along with the amount of electricity demand for 24-hours and IEEE 64-bus test system are depicted in Fig. 10.18.

According to Fig. 10.18, we have the following points:

1. Because solar radiation varies during a day, the amount of it in each hour is defined as a parameter.
2. In some cases, we face a large amount of data such as the amount of energy demand for each hour in the large-scale system. In such conditions, inputting the value of energy demand for each hour similar to the solar radiation will be very time-consuming work. To avoid this, GAMS enables its users to use the possibility of linking with Excel

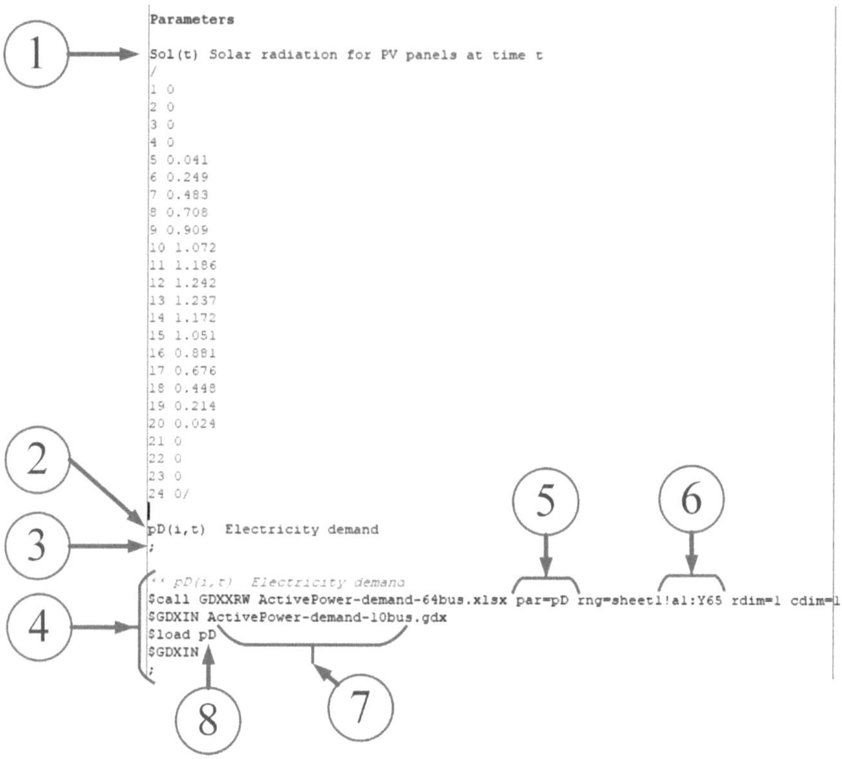

FIGURE 10.18 The example for the Parameter command in GAMS.

files for facilitating the inputting of data. Herein, we have inputted the electricity demand using the linking code between GAMS and Excel.
3. After ending each command in GAMS, we should not forget to insert a semicolon.
4. Presents the set of linking code between the GAMS and Excel for entering data from Excel to GAMS.
5. The name of the parameter should be written here.
6. Presents the address of cells in Excel that we want to input their data in GAMS.
7. The name of the Excel file should be written here.
8. The name of the parameter should also be written here.

10.8.2.3 Table

The Table format can be used for entering multi-dimensional data with respect to at least 2 sets in GAMS. All Table elements have zero amounts in default and we can enter the intended amount to each Table according to the Table general format illustrated in Fig. 10.19.

```
Table Item name (Set1, Set2, ...)  Descriptive text

                  Set2_element1   Set2_element2  ...
Set1_element1        value11          value12
Set1_element2        value21          value22
      ⋮
;
```

FIGURE 10.19 The definition of the Table command in GAMS.

```
(1) → Table  R(i,j)   Rsistance matrix for IEEE 10 bus test system
           1      2       3      4       5       6       7       8       9       10
      1    0      0.1233  0      0       0       0       0       0       0       0
      2    0      0       0.014  0       0       0       0       0       0       0
      3    0      0       0      0.7463  0       0       0       0       0       0
      4    0      0       0      0       0.6984  0       0       0       0       0
      5    0      0       0      0       0       1.9831  0       0       0       0
      6    0      0       0      0       0       0       0.9053  0       0       0
      7    0      0       0      0       0       0       0       2.0552  0       0
      8    0      0       0      0       0       0       0       0       4.7953  0
      9    0      0       0      0       0       0       0       0       0       5.3434
      10   0      0       0      0       0       0       0       0       0       0
(2) → ;
```

FIGURE 10.20 The example for the Table command in GAMS.

As an example, the amount of resistance in the lines of the IEEE 10-bus test system are shown in Fig. 10.20.

Given Fig. 10.20, we have the following points:

1. Presents the resistance matrix for the IEEE 10-bus system that is entered using the Table command in GAMS.
2. Reminding us to don't forget to use the semicolon at the end of each command.

10.8.3 Variable

Unlike parameters, the amount of variables in the mathematical models is not certain from the beginning and will be obtained after solving the problem. Typically, GAMS looking forwards to find optimal solutions for the variables by solving the problem. Variables can be defined over 0 to 20 sets that indicate the possibility of associating one variable with a single case or numerous individual variables. The general format for variables in GAMS is demonstrated in Fig. 10.21.

10.8.3.1 Type of variables in GAMS

In GAMS, there are several types of variables that can be used for various purposes. The different types of variable along with their allowable range for changing are listed in Table 10.5.

Variable

First variable name — Optional text — /Optional value for the attribute/

Second variable name (Set dependency) — Optional text — /Optional values for the attributes/

...

;

FIGURE 10.21 The definition of the Variable command in GAMS.

TABLE 10.5 Different types of variable in GAMS [13].

Type of variable	Permissible range for changing variable
Free	$-\infty < x < +\infty$
Positive	$0 < x < +\infty$
Negative	$-\infty < x < 0$
Binary	0 or 1
Integer	0, 1, 2, ÅÅÅ, 100 (as default)

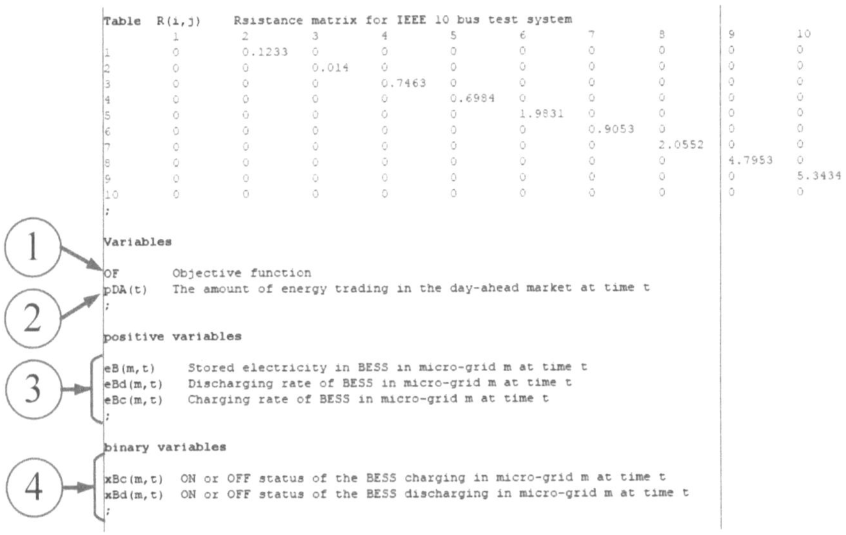

```
Table  R(i,j)    Rsistance matrix for IEEE 10 bus test system
        1        2        3        4        5        6        7        8        9        10
1       0        0.1233   0        0        0        0        0        0        0        0
2       0        0        0.014    0        0        0        0        0        0        0
3       0        0        0        0.7463   0        0        0        0        0        0
4       0        0        0        0        0.6984   0        0        0        0        0
5       0        0        0        0        0        1.9831   0        0        0        0
6       0        0        0        0        0        0        0.9053   0        0        0
7       0        0        0        0        0        0        0        2.0552   0        0
8       0        0        0        0        0        0        0        0        4.7953   0
9       0        0        0        0        0        0        0        0        0        5.3434
10      0        0        0        0        0        0        0        0        0        0
;

Variables

OF        Objective function
pDA(t)    The amount of energy trading in the day-ahead market at time t
;

positive variables

eB(m,t)     Stored electricity in BESS in micro-grid m at time t
eBd(m,t)    Discharging rate of BESS in micro-grid m at time t
eBc(m,t)    Charging rate of BESS in micro-grid m at time t
;

binary variables

xBc(m,t)   ON or OFF status of the BESS charging in micro-grid m at time t
xBd(m,t)   ON or OFF status of the BESS discharging in micro-grid m at time t
;
```


FIGURE 10.22 The example for the Variable command in GAMS.

It should be noted that, in GAMS, the variable that is considered to be optimized as an objective function should be the free variable. The example of using various types of variables in the GAMS environment is demonstrated in Fig. 10.22 considering the mathematical models provided in Chapters 6 and 7.

TABLE 10.6 Suffix of variables in GAMS.

Suffix of variable	Description
.L	Value (level)
.UP	Upper boundary
.LO	Lower boundary
.FX	Fixed value
.m	Dual value

Given Fig. 10.22, we have the following points:

1. Presents the variable that is considered as the objective function and should be defined as the free variable in GAMS.
2. Presents the amount of energy trading in the day-ahead market at time t that is defined as a free variable because its amount can be either positive or negative in the scheduling problem.
3. Presents some of the positive variables like the charging rate of BESS that only can have a positive amount in the problem.
4. Presents some of the binary variables such as the charging state of BESS in the system that only can be 0 (on) or 1 (off).

Each variable can have an appropriate suffix in the model that all of them are tabulated in Table 10.6.

As an example, Fig. 10.23 shows the application of various suffixes for variables in GAMS.

Given Fig. 10.23, we have the following points:

1. We have defined voltage angle as the free variable because its amount can be positive or negative after solving the problem.
2. Because the amount of voltage angle is zero for the slack bus, we have fixed its amount for bus 1 using the suffix ".fx".
3. The amount of voltage angle in all buses should not be greater than 3.14 that is limited using the suffix ".up".
4. The amount of voltage angle in all buses should not be less than −3.14 that is limited using the suffix ".lo".
5. We have assumed the maximum amount of energy sold to the day-ahead market can be up to 3000 kW.
6. We have assumed the maximum amount of energy purchased from the day-ahead market can be up to −3000 kW.

10.8.4 Eq

After orderly defining the Sets, Scalars, Parameters, Tables, and Variables, now it is time to define the model's Eqs.. Eqs. are taken into account

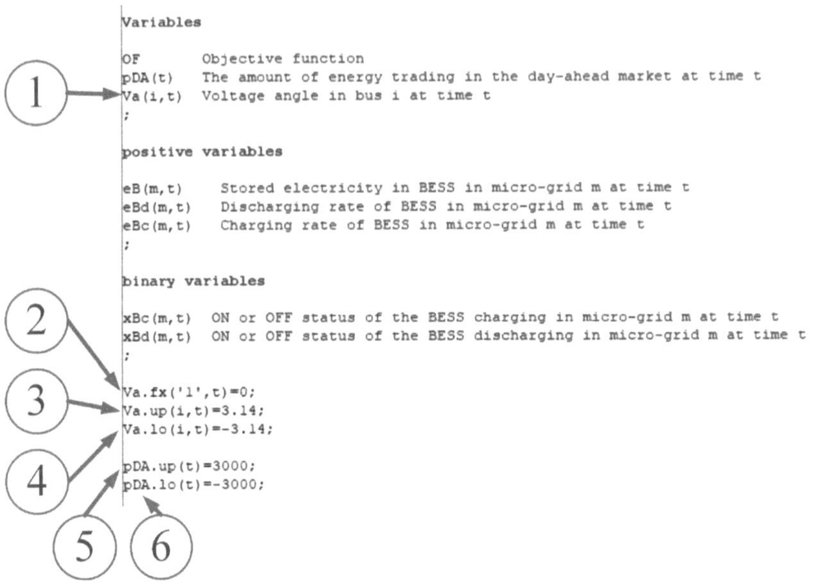

```
Variables

OF        Objective function
pDA(t)    The amount of energy trading in the day-ahead market at time t
Va(i,t)   Voltage angle in bus i at time t
;

positive variables

eB(m,t)     Stored electricity in BESS in micro-grid m at time t
eBd(m,t)    Discharging rate of BESS in micro-grid m at time t
eBc(m,t)    Charging rate of BESS in micro-grid m at time t
;

binary variables

xBc(m,t)   ON or OFF status of the BESS charging in micro-grid m at time t
xBd(m,t)   ON or OFF status of the BESS discharging in micro-grid m at time t
;

Va.fx('1',t)=0;
Va.up(i,t)=3.14;
Va.lo(i,t)=-3.14;

pDA.up(t)=3000;
pDA.lo(t)=-3000;
```


FIGURE 10.23 The example of the variables suffix in GAMS.

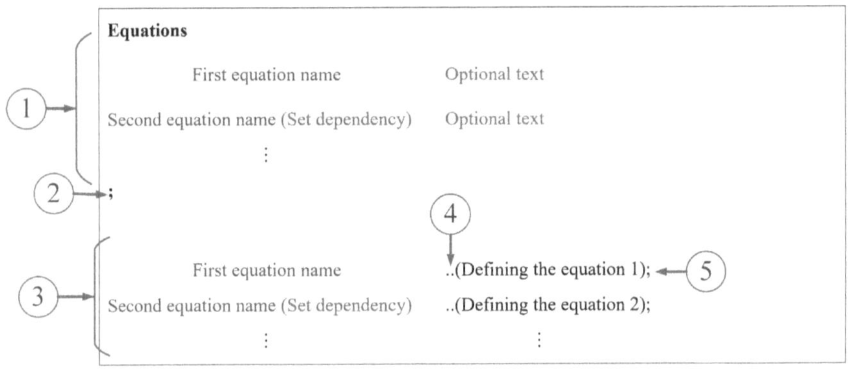

FIGURE 10.24 The format of defining Eqs. in GAMS.

as the core of the structure of each model in GAMS. The general format for defining Eqs. in GAMS is shown in Fig. 10.24, which consists of two steps.

Given Fig. 10.24, we have the following points:

1. Presents the first step of defining Eqs. in GAMS. In this step, we need to define the name of all Eqs. along with their related descriptions (these descriptions are optional).
2. After defining the name of all Eqs., we need to end the first step of Eqs. by inserting a semicolon. Inserting this semicolon is essential for avoiding the related error after running the program in GAMS.

TABLE 10.7 Some of the mathematical signs in GAMS [13].

Format in GAMS	Typical format in Eqs.	Description
=E=	=	The left-hand side of the Eq. is equal to the right-hand side.
=L=	≤	The right-hand side of the Eq. is greater than or equal to the left-hand side of it.
=G=	≥	The right-hand side of the Eq. is less than or equal to the left-hand side of it.

3. Presents the second part of defining Eqs.. In this step, we need to write the name of the Eqs. in accordance with the first step and then insert two dots for defining the related Eqs..
4. Inserting at least one space and two dots in front of each Eq. is necessary for defining the related Eq.
5. After defining the Eq. in each line (row), we need to separate all lines from each other by inserting a semicolon at the end of each Eq.

In Eqs., we usually face with the signs of "equal or =" and "greater or less than (≤ ≥)". The equivalents of these signs in GAMS are listed in Table 10.7.

Herein, we have selected three sample systems between the clean energy production systems, controllable hybrid systems, and energy storage systems (one system from each group) to indicate their equivalent models in GAMS. For this aim, the mathematical models of PV panels, combined cooling, heating and power (CCHP) unit [15], and liner version of BESS [16] are implemented in GAMS according to Fig. 10.25.

According to Fig. 10.25, we have the following points:

1. Presents the first step of defining Eqs. in GAMS, which in this example, it includes the Eqs. of the PV panel, CCHP unit, and BESS.
2. The name of Eqs. can be written in one line (row) instead of writing each name in one row aiming to avoid making a large model. For this aim, each Eq.'s name should be separated from each other by using a comma.
3. Using a semicolon is essential for separating the first and second parts of the Eqs..
4. Presents the second part of defining Eqs. that the name of all Eqs. is written again with the aim of defining their mathematical formula.
5. Presents the Eq. of the PV panel model.
6. Presents the Eqs. of the CCHP model.
7. Presents the Eqs. of the BESS model.
8. Presents this Eq. only needs to be satisfied for $t = 1$.

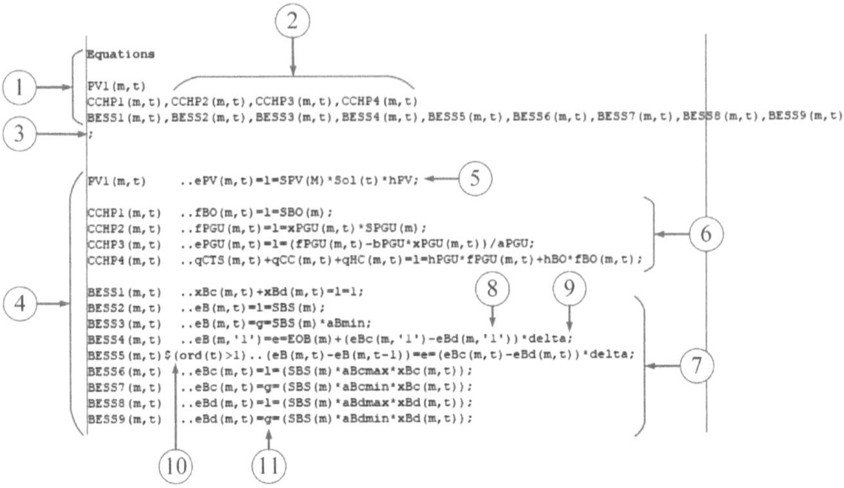

FIGURE 10.25 The example for the Eq. command in GAMS.

Model

Model name 1	Optional text	/ Equations that we want to solve in Model 1/
Model name 2	Optional text	/ Equations that we want to solve in Model 2/
⋮	⋮	⋮
;		

FIGURE 10.26 The general form of the Model in GAMS.

9. At the end of each Eq., we need to insert a semicolon.
10. This is the "ord" command that is fully described in the next sections of this chapter.
11. According to Table 10.7, we need to use "=g=" instead of ≥ in GAMS.

It should be noted that, before writing Eqs., we need to orderly define all the related Sets, Scalars, Parameters, Tables, and Variables.

10.8.5 Model

Models are the collection of Eqs. with different variables and are taken into account as the objects that are considered to be solved in GAMS. The general form of the Model is depicted in Fig. 10.26.

According to the general form of the Model, we can have more than one model in each GAMS program. The type of each problem in GAMS is typically determined based on the type of Eqs. in each model. For example, if we have nonlinear Eqs. in our model, the type of problem would be

TABLE 10.8 Some of the most common solution procedures in GAMS [13].

Solution procedure	Description
LP	The linear programming (LP) does not contain nonlinear or discrete (integer and binary) variables.
MIP	The mixed-integer linear programming (MIP) can only contain linear and discrete (integer and binary) variables.
NLP	The nonlinear programming (NLP) can only contain nonlinear variables without discrete (integer and binary) variables.
MINLP	The mixed-integer nonlinear programming (MINLP) simultaneously contain nonlinear and discrete (integer and binary) variables.
DNLP	The nonlinear programming with discontinuous derivatives (DNLP) is a description of nonlinear programming with discontinuous derivatives.

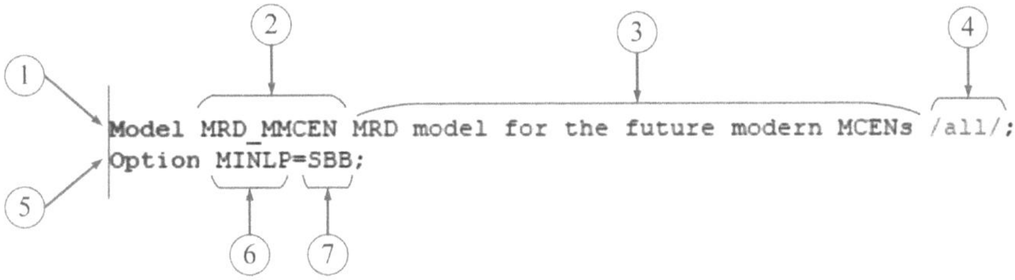

FIGURE 10.27 The example for the Eq. command in GAMS.

NLP. Some of the most common types of problems in GAMS are listed in Table 10.8.

The next line after the Model line in GAMS can be typically the Option line. Option command allows the user to directly determine the type of solver in accordance with the type of problem. For example, if our problem is MINLP, we can use several solvers in GAMS for solving it. By using the Option command, we can easily determine that for example, we want to use an SBB solver for solving our MINLP problem. The example for Model and Option commands in GAMS is demonstrated in Fig. 10.27.

According to Fig. 10.27, we have the following points:

1. Presents the Model line with its various parts.
2. Presents the name of the model.
3. Presents the descriptions for the model. The existence of this part is optional.
4. Presents the Eqs. that are considered in solving the process of the problem. In this part of the model, we can identify which Eqs. need to be intended in the problem-solving process.

Solve	The name of the Model	The keyword "using"	An available solution procedure	The keyword "maximizing" or "minimizing"	The name of the variable to be optimized

FIGURE 10.28 The general form of the Solve command in GAMS.

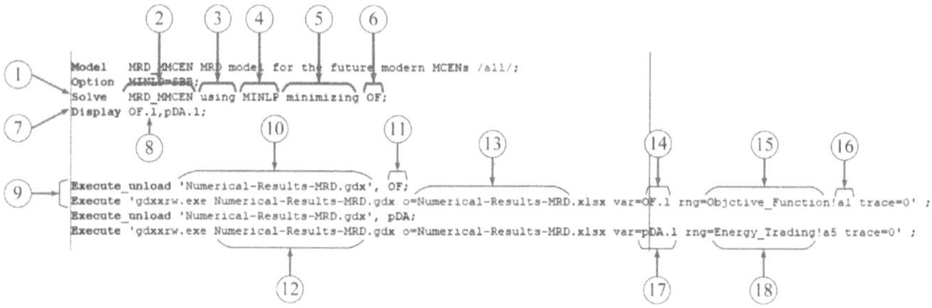

FIGURE 10.29 The example for the Solve and Display commands in GAMS.

5. Presents the Option line with its different parts.
6. Presents the type of problem that here is the MINLP problem.
7. Presents the type of solver that we want to be considered in GAMS for solving the problem.

10.8.6 Solve and display

Defining a Solve statement is the last step in developing one Model in GAMS. The general format for defining a Solve statement is demonstrated in Fig. 10.28.

After writing the Solve statement, we can use the Display command for determining those variables that we want to receive their amounts after solving the problem. Herein, we can use suffixes ".L" and ".M" for respectively achieving the primal and dual amounts of each variable. The example for the application of Solve and Display commands in GAMS is illustrated in Fig. 10.29.

According to Fig. 10.29, we have the following points:

1. Presents the Solve statement.
2. Presents the name of the model.
3. We need to use the "using" keyword after the name of the model in the Solve statement.
4. Presents the type of mathematical problem.
5. We need to determine that we want to maximize a certain variable or minimize it in the developed Model.
6. Presents the name of the variable that is aimed to be maximized or minimized in GAMS.

7. Presents the Display line.
8. We can write any variables of the Model in the Display line by adding the suffix ".L" at the end of them for receiving their primal amounts from GAMS. If we want to extract the dual amount of each variable, we should insert the suffix ".M" instead of the suffix ".L".
9. In most cases, we need to extract different results with a large volume. In such conditions, directly extracting all results from the typical GAMS environment for evaluation or other purposes is very time-consuming work. Herein, we can use this code for directly storing the desired results in the predetermined Excel file. For this aim, at first, we need to create an empty Excel file in the folder that we already created the GAMS file in it. Then, we use the appropriate name for the Excel file to easily find it after solving the problem. For example, we have created the Excel file with the name "Numerical-Results-MRD". Afterward, we can use this code at the end of the Model for directly extracting the results from GAMS and storing them in the Excel file.
10. Presents the name of the Excel file.
11. Presents the name of the variable that we want to store its amount in the Excel file.
12. Presents the name of the Excel file.
13. Presents the name of the Excel file.
14. Presents the name of the variable that we want to store its amount in the Excel file.
15. We can easily store the different results in the various sheet of the Excel file. This work helps us to better categorize the results. This part of the code presents the name of the sheet in the Excel file.
16. By using this part of the code, we can easily determine in which line of the Excel file we want to store the amount of the related variable.
17. Presents the name of the other variable that we want to store its amount in the Excel file.
18. Presents the name of another sheet in the Excel file.

After completing all the aforementioned steps, we can run the program in GAMS. It should be noted that all GAMS solvers are deeply described in [17].

10.9 Some special orders in GAMS

In GAMS, there are several applicable orders that we can easily access them in the Help tab of the GAMS software. Describing all of them is not in the scope of this chapter, but we have brought some of the most common of them in the following subsections.

```
LOOP ((Set1, Set2,...) [$(condition)],
Statement or Statements to Execute;
);
```

FIGURE 10.30 The general form of the Loop statement in GAMS.

```
IF (Condition,
        Statements;
{ElseIf Condition,
        Statements;}
 [Else
        Statements;]
);
```

FIGURE 10.31 The general form of the If-Elseif-Else statement in GAMS.

10.9.1 Loop statement

The Loop statement enables the user to execute a group of Eqs. and statements for each element of a set. Indeed, by using the Loop statement, we can easily obtain the solution of the objective function for each element in a set. The Loop statement can be written according to Fig. 10.30.

It should be noted that the Sets, Scalars, Parameters, Tables, Acronyms, For, Variables, Eqs. cannot be placed inside the Loop statement.

10.9.2 If-Elseif-Else statement

In order to transfer from one operator to another one, one of the useful statements is the IF-ELSE operator. In GAMS, we can write the IF-ELSE statement using the illustrated general form in Fig. 10.31.

It should be noted that the Sets, Scalars, Parameters, Tables, Acronyms, For, Variables, Eqs. cannot be placed inside the If-Elseif-Else statement.

10.9.3 While statement

The While statement enables the user to repeatedly execute a block of Eqs. and statements until a logical condition is satisfied. The general form of the While statement is shown in Fig. 10.32.

It should be noted that the Sets, Scalars, Parameters, Tables, Acronyms, For, Variables, Eqs. cannot be placed inside the While statement.

```
WHILE (Condition,
        Statement or Statements to Execute;
);
```

FIGURE 10.32 The general form of the While statement in GAMS.

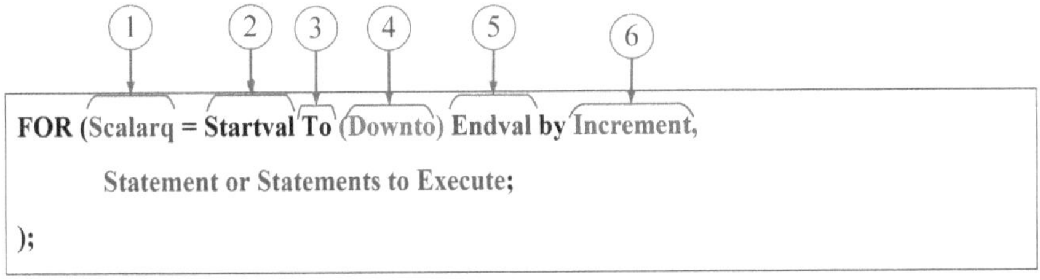

FIGURE 10.33 The general form of the For statement in GAMS.

10.9.4 For statement

The For statement enables the user to repeatedly execute a block of Eqs. and statements over successively varied values of a scalar. The general form of the For statement is depicted in Fig. 10.33.

According to Fig. 10.33, we have the following points:

1. It is a scalar.
2. It is the scalar or constant that the Scalarq will begin from its value.
3. It indicates that the increment will be conducted by GAMS until Scalarq gets equal to or larger than end.
4. It indicates that the subtract increment will be conducted by GAMS until Scalarq gets equal to or smaller than end.
5. It is scalar or constant giving the value that will result in statement termination when Scalarq equals or passes it.
6. It is a positive scalar or constant that is optional and defaults to one.

It should be noted that the Sets, Scalars, Parameters, Tables, Acronyms, For, Variables, Eqs. cannot be placed inside the For statement.

10.9.5 Repeat statement

The Repeat statement enables the user to use it for repeatedly executing a block of Eqs. and statements over and over until a logical condition is satisfied. The general form of the Repeat statement is depicted in Fig. 10.34.

It should be noted that the Sets, Scalars, Parameters, Tables, Acronyms, For, Variables, Eqs. cannot be placed inside the Repeat statement.

In writing the statement of the different operators, we may face with some numerical relationship operators that are listed in Table 10.9.

```
REPEAT (
        Statement or Statements to Execute;
UNTIL logical condition is true
);
```

FIGURE 10.34 The general form of the Repeat statement in GAMS.

TABLE 10.9 Some numerical relationship operators in GAMS [13].

Operator		Meaning
Gt	>	Strictly greater than
Ge	>=	Greater than or equal to
Eq	=	Equal to
Ne	< >	Not equal to
Lt	<	Strictly less than
Le	<=	Less than-or-equal to

TABLE 10.10 Some logical operators in GAMS [13].

Operator	Meaning
And	And
Or	Inclusive or
Not	Not
Xor	Exclusive or $(\bar{C}.D + C.\bar{D})$

Moreover, some logical useful operators in GASM are tabulated in Table 10.10.

10.9.6 ORD operator

The ORD operator makes the amount of the Parameter equal to the related index position in a set. The following example in Fig. 10.35 clearly shows its application in GAMS.

10.9.7 CARD operator

The CARD operator makes the amount of the Parameter equal to the number of elements in the related set. The following example in Fig. 10.36 clearly illustrates its application in GAMS.

```
SET
t  Time periods  /15*24/;
Parameter
A (t) ;
A (t) = ORD(t) ;

Results are as follows:
A (15) = 1
A (16) = 2
   ⋮
A (16) = 10
```

FIGURE 10.35 The example of the ORD operator application in GAMS.

```
SET
t  Time periods  /15*24/;
Parameter
A ;
A = CARD (t) ;

Result is as follows:
A = 10
```

FIGURE 10.36 The example of the CARD operator application in GAMS.

10.10 Summary

This chapter is focused on providing useful and time-saving materials regarding the GAMS application in the MCENs. In the energy sector, the mathematical models proposed for each system or network need to be simulated using the appropriate software before implementing in the practice. The ever-increasing of optimization problems in MCENs has drastically created the need for optimization tools for effectively analyzing the different types of problems in the system. In this regard, GAMS software with several effective solvers is specialized for various optimization problems. Due to the undeniable capability of the GAMS in solving diverse optimization problems, its usage in future modern MCENs will be inevitable. This was the main motivation for us to briefly indicate how we can learn to work with GAMS and solve the optimization problems of MCENs in a short time after a deep investigation of them in terms of conceptual and mathematically in the previous chapters. Hence, we attempted in this chapter to applicably cover all steps that we require to develop a Model and solve it in GAMS. In each step, useful examples are provided to make understanding the related concept or application of a special command easier. In this regard, almost all of the examples are developed based on the mathematical models provided in chapters 6 and 7 to facilitate the implementation of similar models in GAMS.

10.11 Questions

The careful analysing whole of this chapter is expected to enable readers to effectively evaluate the following questions.

1) What is GAMS in the optimization world?
2) Why the usage of GAMS is necessary for MCENs?
3) What are the distinguishing features of the GAMS in optimizing the different problems?
4) What are the common terms, symbols, and reserved words in GAMS and how can we use them in GAMS?
5) What are the common mathematical functions in GAMS and how can we use them in GAMS?
6) What are the main GAMS commands and how can we use them in GAMS?
7) What are the main steps for developing one Model in GAMS?
8) From how many ways we can input the different types of data in GAMS?
9) What are the general mathematical signs in GAMS and how they can be used in GAMS?
10) What is the structure of each Model in GAMS?
11) What are the most common solution procedures in GAMS?
12) What is the structure of the Solve statement in each Model of GAMS?
13) What are the special orders in GAMS and how can we use them in GAMS?
14) What are the numerical relationship operators in GAMS and how can we use them in GAMS?
15) What are the logical operators in GAMS and how can we use them in GAMS?

10.12 Suggestions

Given the topic of this chapter, the following suggestions are provided for the readers assisting them to pursue related research clues.

1) Designing the full option optimization software is one of the effective steps in solving the complex optimization problems of future modern MCENs.

2) Employing the GAMS software can facilitate the solving process of the different optimization problems in future modern MCENs due to having various solvers with useful features.
3) Upgrading GAMS solvers can make the strongly facing the new types of optimization problems easier.
4) Developing the GAMS and equipping it with the new commands can improve the solving process of optimization problems.

10.13 Future trends

Future trends can give a proper overview for the readers to form their related works and develop effective solutions for remaining gaps, which are listed as follows for this chapter.

1) Designing the new optimization software for facilitating the solving process of optimization problems with reliable solutions.
2) Upgrading GAMS solvers for expanding its application in the optimization world.
3) Developing new commands for facilitating the building process of various mathematical models in GAMS.

References

[1] M. Daneshvar, E. Babaei, Exchange Market Algorithm for Multiple DG Placement and Sizing in a Radial Distribution System, Journal of Energy Management and Technology 2 (1) (2018) 54–65.
[2] H. Jahangir, S.S. Gougheri, B. Vatandoust, M.A. Golkar, A. Ahmadian, A. Hajizadeh, Plug-in Electric Vehicle Behavior Modeling in Energy Market: a Novel Deep Learning-Based Approach with Clustering Technique, IEEE Trans Smart Grid (2020).
[3] M. Daneshvar, B. Mohammadi-ivatloo, S. Asadi, S. Galvani, Short Term Optimal Hydro-Thermal Scheduling of the Transmission System Equipped with Pumped Storage in the Competitive Environment, Majlesi Journal of Electrical Engineering 14 (1) (2020) 77–84.
[4] M. Daneshvar, B.M. Ivatloo, M. Abapour, S. Asadi, R. Khanjani, Distributionally Robust Chance Constrained Transactive Energy Framework for Coupled Electrical and Gas Microgrids, IEEE Transactions on Industrial Electronics (2020).
[5] M. Daneshvar, H. Eskandari, A.B. Sirous, R. Esmaeilzadeh, A novel techno-economic risk-averse strategy for optimal scheduling of renewable-based industrial microgrid, Sustainable Cities and Society 70 (2021) 102879.
[6] V. Robichaud, An introduction to GAMS, ed, 2010.
[7] M. Daneshvar, B. Mohammadi-ivatloo, S. Asadi, M. Abapour, A. Anvari-Moghaddam, A Transactive Energy Management Framework for Regional Network of Microgrids, in: 2019 International Conference on Smart Energy Systems and Technologies (SEST), 2019, pp. 1–6.
[8] A. Soroudi, Power System Optimization Modeling in GAMS, Springer, 2017.
[9] (23 Jun, 2020). *An Introduction to GAMS*. Available: https://www.gams.com/products/introduction/
[10] M. Daneshvar, B. Mohammadi-Ivatloo, A. Anvari Moghaddam, Optimal Energy Trading Strategy for Proactive DISCO considering Demand Response Programs in the

distribution networks, in: CIGRE Symposium Aalborg 2019, CIGRE (International Council on Large Electric Systems, 2019.
[11] M. Daneshvar, M. Abapour, B. Mohammadi-ivatloo, S. Asadi, Impact of optimal DG placement and sizing on power reliability and voltage profile of radial distribution networks, Majlesi Journal of Electrical Engineering 13 (2) (2019) 91–102.
[12] (23 Jun, 2020). Download GAMS Release 31.2.0 (June 19, 2020). Available: https://www.gams.com/download/
[13] M.B. Abaee, M.S. Ghazizadeh, Generalized Algebraic Modeling System (GAMS), PWUT Spring, 2010.
[14] B. Ulanicki, P. Bounds, J. Rance, Using a GAMS modelling environment to solve network scheduling problems, Measurement and Control 32 (4) (1999) 110–115.
[15] M. Daneshvar, B. Mohammadi-Ivatloo, K. Zare, M. Abapour, S. Asadi, A. Anvari-Moghaddam, Chance-constrained scheduling of hybrid microgrids under transactive energy control, Int. J. Energy Res. 45 (7) (2021) 10173–10190.
[16] M. Daneshvar, B. Mohammadi-Ivatloo, K. Zare, S. Asadi, Two-stage robust stochastic model scheduling for transactive energy based renewable microgrids, IEEE Transactions on Industrial Informatics 16 (11) (2020) 6857–6867.
[17] G.D.C.GAMS: the Solver Manuals, GAMS Development Corporation, 2010.

Index

Page numbers followed by "*f*" and "*t*" indicate, figures and tables respectively.

FSC
www.fsc.org
MIX
Papier | Fördert gute Waldnutzung
FSC® C083411

Kolibri 360 GmbH
Ferdinand-Jühlke-Straße 7
99095 Erfurt, Deutschland
produktsicherheit@kolibri360.de